AF613709

Universitext

Dear MyCopy Customer,

This Springer book is a monochrome print version of the eBook to which your library gives you access via SpringerLink. It is available to you at a subsidized price since your library subscribes to at least one Springer eBook subject collection.

Please note that MyCopy books are only offered to library patrons with access to at least one Springer eBook subject collection. MyCopy books are strictly for individual use only.

You may cite this book by referencing the bibliographic data and/or the DOI (Digital Object Identifier) found in the front matter. This book is an exact but monochrome copy of the print version of the eBook on SpringerLink.

Springer Science+Business Media, LLC

Universitext

Editors (North America): S. Axler, F.W. Gehring, and K.A. Ribet

Aksoy/Khamsi: Nonstandard Methods in Fixed Point Theory
Andersson: Topics in Complex Analysis
Aupetit: A Primer on Spectral Theory
Bachman/Narici/Beckenstein: Fourier and Wavelet Analysis
Badescu: Algebraic Surfaces
Balakrishnan/Ranganathan: A Textbook of Graph Theory
Balser: Formal Power Series and Linear Systems of Meromorphic Ordinary Differential Equations
Bapat: Linear Algebra and Linear Models (2nd ed.)
Berberian: Fundamentals of Real Analysis
Boltyanskii/Efremovich: Intuitive Combinatorial Topology. (Shenitzer, trans.)
Booss/Bleecker: Topology and Analysis
Borkar: Probability Theory: An Advanced Course
Böttcher/Silbermann: Introduction to Large Truncated Toeplitz Matrices
Carleson/Gamelin: Complex Dynamics
Cecil: Lie Sphere Geometry: With Applications to Submanifolds
Chae: Lebesgue Integration (2nd ed.)
Charlap: Bieberbach Groups and Flat Manifolds
Chern: Complex Manifolds Without Potential Theory
Cohn: A Classical Invitation to Algebraic Numbers and Class Fields
Curtis: Abstract Linear Algebra
Curtis: Matrix Groups
Debarre: Higher-Dimensional Algebraic Geometry
DiBenedetto: Degenerate Parabolic Equations
Dimca: Singularities and Topology of Hypersurfaces
Edwards: A Formal Background to Mathematics I a/b
Edwards: A Formal Background to Mathematics II a/b
Farenick: Algebras of Linear Transformations
Foulds: Graph Theory Applications
Friedman: Algebraic Surfaces and Holomorphic Vector Bundles
Fuhrmann: A Polynomial Approach to Linear Algebra
Gardiner: A First Course in Group Theory
Gårding/Tambour: Algebra for Computer Science
Goldblatt: Orthogonality and Spacetime Geometry
Gustafson/Rao: Numerical Range: The Field of Values of Linear Operators and Matrices
Hahn: Quadratic Algebras, Clifford Algebras, and Arithmetic Witt Groups
Heinonen: Lectures on Analysis on Metric Spaces
Holmgren: A First Course in Discrete Dynamical Systems
Howe/Tan: Non-Abelian Harmonic Analysis: Applications of $SL(2, \mathbb{R})$
Howes: Modern Analysis and Topology
Hsieh/Sibuya: Basic Theory of Ordinary Differential Equations
Humi/Miller: Second Course in Ordinary Differential Equations
Hurwitz/Kritikos: Lectures on Number Theory

(continued after index)

André Martinez

An Introduction to Semiclassical and Microlocal Analysis

Springer

André Martinez
Department of Mathematics
University of Bologna
Bologna, 40127 Italy
martinez@dm.unibo.it

Mathematics Subject Classification (2000): 35Sxx, 58Jxx, 81Q20

Library of Congress Cataloging-in-Publication Data
Martinez, André.
An introduction to semiclassical and microlocal analysis / André Martinez.
p. cm. — (Universitext)
Includes bibliographical references and index.
DOI 10.1007/978-1-4757-4495-8

1. Microlocal analysis. 2. Quantum theory. I. Title. II. Series.
QC20.7.M53 M37 2001
530.15′57242—dc21 2001049263

Printed on acid-free paper.

Production managed by Terry Kornak; manufacturing supervised by Jeffrey Taub.
Photocomposed copy prepared by the author.

9 8 7 6 5 4 3 2 1

www.springer.com/mycopy

Preface

The following lecture notes correspond to a course taught for several years, first at the University of Paris-Nord (France) and then at the University of Bologna (Italy). They are mainly addressed to nonspecialists in the subject, and their purpose is to present in a pedagogical way most of the techniques used in the microlocal treatment of semiclassical problems coming from quantum physics. Both the standard C^∞ pseudodifferential calculus and the analytic microlocal analysis are developed, in a context that remains intentionally global so that only the relevant difficulties of the theory are encountered. The main originality lies in the fact that we derive all the main features of analytic microlocal analysis from a single a priori estimate, which turns out to be elementary once the C^∞ pseudodifferential calculus is established.

Various detailed exercises are given at the end of the main chapters, most of them being easily solvable by students. Besides illustrating the main results of the lecture, their aim is also to introduce the reader to various further developments of the theory, such as the functional calculus of pseudodifferential operators, properties of the analytic wave front set, Gevrey classes, the use of coherent states, the notion of semiclassical measures, WKB constructions. Applications to the study of the Schrödinger operator are also discussed in the text, so that they may help the understanding of new notions or general results where they appear by replacing them in the context of quantum mechanics. We invite the reader who wishes to find these applications easily to refer to the index, which we have tried to make as complete as possible.

The prerequisities are essentially reduced to the basic notions of the theory of distributions.

Acknowledgements: Many people have contributed—by their advice, suggestions, corrections, references, remarks, or encouragements—to the presentation of this book. In particular, we address many thanks to:
Christian BROUDER, Sandro GRAFFI, Vincenzo GRECCHI, Bernard HELFFER, Shu NAKAMURA, Didier ROBERT, Akihiro SHIMOMURA, Johannes SJÖSTRAND, Mark SOFRONIU.

We are also very grateful to the three anonymous referees for their constructive remarks and useful references, and to the whole staff of Springer–Verlag New York for the excellent work they did in the editing process of this book.

A special mention to my wife, Vania SORDONI, not only for all the aforementioned reasons, but also for her constant efforts to make my life more pleasant.

André Martinez

Contents

Chapter 1

Introduction

1.1 A Short Review of Classical Mechanics

In this section and in the following one our aim is only to give a slight flavor of the process that has led physicists to change completely their conception of reality by passing from classical mechanics to quantum mechanics. For a much more detailed presentation of this, we refer the reader to the very complete book [Mes], by A. Messiah. Interesting information can also be found in the book [Ro], by D. Robert (which is actually closer to our point of view), and in the classical book [LaLi], by L.D. Landau and E.M. Lifshitz. (Indeed, a large literature—in any language—exists on the foundations of quantum mechanics, and many more references can be found than we can give here.)

In classical mechanics, a particle of mass m is represented by its *position* at time t, that is, by a function $t \mapsto x(t) \in \mathbf{R}^3$. If this particle is submitted to a conservative force field $F = -\nabla V$, then its movement is ruled by Newton's fundamental law:

$$F = m\ddot{x}(t), \tag{1.1.1}$$

or equivalently,

$$\ddot{x}(t) = -\frac{1}{m}\nabla V\left(x(t)\right), \tag{1.1.2}$$

where the dots stand for differentiations with respect to t. If we set $\xi(t) = m\dot{x}(t)$ (the so-called *momentum* or *impulse* of the particle), then (1.1.2) can be rewritten as

$$\begin{cases} \dot{\xi}(t) = -\nabla V(x(t)), \\ \dot{x}(t) = \frac{1}{m}\xi(t), \end{cases} \tag{1.1.3}$$

which is called a system of Hamilton's equations. The curve $t \mapsto (x(t), \xi(t))$ is then called the *phase space trajectory* or *classical trajectory* of the particle, and lies in $\mathbf{R}^3 \times \mathbf{R}^3$, which should be viewed as the product of the space of positions and momenta.

The *(total) energy* of the particle is defined by

$$E = \frac{1}{2m}\xi(t)^2 + V(x(t)). \tag{1.1.4}$$

The main feature of the energy is that it is independent of t: Indeed, one has

$$\frac{d}{dt}\left(\frac{1}{2m}\xi(t)^2 + V(x(t))\right) = m\ddot{x}(t) \cdot \dot{x}(t) + \nabla V(x(t)) \cdot \dot{x}(t) = 0,$$

where the last equality is a direct consequence of (1.1.2). Note that the energy of the particle is just the value at $(x(t), \xi(t))$ of the so-called *energy observable* $\frac{1}{2m}\xi^2 + V(x)$.

More generally, one calls a *classical observable* any real smooth function $a = a(x, \xi)$ defined on the phase space $\mathbf{R}^3 \times \mathbf{R}^3$: Its value at the point $(x(t), \xi(t))$ gives information about the particle at time t. In particular, any physical experiment concerning the particle should lead to quantities that can be described by such values of classical observables.

However, it turns out that in several experiments (such as the photoelectric effect and the diffraction of particles) (see [Mes, Ro]), properties in contradiction to this classical model of mechanics have appeared. The most well known are the facts that the energy of a particle can take values only in a discrete subset of $\mathbf{R}$, and that one cannot know at the same time the precise values of both the position and the momentum of the particle: This is the famous *Heisenberg uncertainty principle*, which asserts that the errors Δx and $\Delta\xi$ made in a measurement of the position and the momentum always satisfy

$$\Delta x \cdot \Delta\xi \geq \frac{h}{4\pi}, \tag{1.1.5}$$

where h is the Planck constant, whose value is approximately 6.6×10^{-34} joules/second.

These observations have led physicists to believe in a kind of double nature of elementary particles, both wavelike and corpuscular. After a first attempt in 1923 by De Broglie [DeB] to include such observations in a mathematical model (the so-called *matter waves*, which generalize to matter the double aspect—wavelike and corpuscular—previously observed in light), it is now commonly admitted that a very general and acceptable model is given by another theory of matter: quantum mechanics, introduced in two equivalent forms around 1925, by M. Born, W. Heisenberg and P. Jordan (the *matrix mechanics*, see, e.g., [BoHeJo]), and by E. Schrödinger (the *wave mechanics*, see [Schr1, Schr2]). About the links between these two presentations, one may consult, e.g., [VdW] and references therein.

1.2 Basic Notions of Quantum Mechanics

In quantum mechanics, a particle is described by a function $\mathbf{R}\times\mathbf{R}^3 \ni (t,x) \mapsto \psi(t,x) \in \mathbf{C}$, which is called the *wave function* of the particle. The wave function must be such that for any $t \in \mathbf{R}$, the function $\psi_t : \ x \mapsto \psi(t,x)$ belongs to $L^2(\mathbf{R}^3)$, and $\|\psi_t\|_{L^2(\mathbf{R}^3)} = 1$. The function ψ_t is called the *state* of the particle at time t.

The natural interpretation attached to $\psi(t,x)$ is to view $|\psi(t,x)|^2$ as a density of probability: It describes the probability of presence of the particle at the point x at time t.

The *average position* of the particle at time t is defined in a natural way as the quantity

$$\langle x\rangle_{\psi_t} := \langle x\psi_t, \psi_t\rangle_{L^2(\mathbf{R}^3)} = \left(\langle x_j\psi_t, \psi_t\rangle_{L^2(\mathbf{R}^3)}\right)_{j=1,2,3}. \tag{1.2.1}$$

An *average impulse* can also be defined, but its understanding requires an analogy with a *plane wave* given in optics by a function of the type

$$\varphi(t,x) = Ae^{i(k\cdot x-\omega t)}, \tag{1.2.2}$$

where $\nu := \dfrac{\omega}{2\pi}$ represents the frequency, and $k \in \mathbf{R}^3$ is called the *wave vector*. The wave propagates along the direction of k, in the sense that φ is independent of x on any plane $\{x \cdot k = \text{constant}\}$. As a consequence, it is natural that any acceptable definition of the impulse of such a wave must satisfy

$$\xi = \alpha k$$

for some positive constant α. Actually, the so-called De Broglie relation (derived from considerations on free wavepackets; see, e.g., [Mes]), gives

$$\xi = \hbar k, \tag{1.2.3}$$

where $\hbar := \dfrac{h}{2\pi}$ is the reduced Planck constant. Using (1.2.2) and (1.2.3) we get in particular:

$$\xi = \frac{\hbar}{i}\left[\nabla_x \varphi(t,x)\right] \overline{\varphi(t,x)}/|A|^2. \tag{1.2.4}$$

It is precisely relation (1.2.4) which provides a way to define by analogy the average impulse of the quantum particle described by $\psi(t,x)$. Viewing $|A|^2$ in (1.2.4) as a normalization factor, one sets

$$\langle \xi \rangle_{\psi_t} := \left\langle \frac{\hbar}{i}\nabla \psi_t, \psi_t \right\rangle_{L^2(\mathbf{R}^3)} = \left(\left\langle \frac{\hbar}{i}\frac{\partial \psi_t}{\partial x_j}, \psi_t \right\rangle_{L^2(\mathbf{R}^3)} \right)_{j=1,2,3} \tag{1.2.5}$$

whenever it is defined (e.g., if $\psi_t \in H^1(\mathbf{R}^3)$, the usual Sobolev space).

At this point it is useful to make a connection (which will be essential in the sequel) between (1.2.5) and an equivalent way of writing $\langle \xi \rangle_{\psi_t}$ by using the so-called *$\hbar$-Fourier transform* of ψ_t:

$$\mathcal{F}_\hbar \psi_t(\xi) := \widehat{\psi}_t(\xi) := \frac{1}{(2\pi\hbar)^{3/2}} \int e^{-ix\xi/\hbar} \psi_t(x)dx. \tag{1.2.6}$$

The $\hbar$-Fourier transform $\mathcal{F}_\hbar$ is an isometry of $L^2(\mathbf{R}^3)$, and if $\psi_t \in H^1(\mathbf{R}^3)$, one has the relation

$$\mathcal{F}_\hbar \left(\frac{\hbar}{i}\nabla \psi_t \right)(\xi) = \xi \mathcal{F}_\hbar \psi_t(\xi).$$

As a consequence, (1.2.5) can be rewritten as

$$\langle \xi \rangle_{\psi_t} = \left\langle \xi \widehat{\psi}_t, \widehat{\psi}_t \right\rangle_{L^2(\mathbf{R}^3)}, \tag{1.2.7}$$

so that it assumes a form more similar to (1.2.1).

Here we observe that the "wave–corpuscule" duality of a quantum particle is mathematically given by the correspondence between $\psi_t(x)$ and $\widehat{\psi}_t(\xi)$ via the $\hbar$-Fourier transform, in the sense that the average impulse of ψ_t equals the average position of $\widehat{\psi}_t$ and vice versa. Moreover, the classical quantities

x and ξ can be considered in quantum mechanics only via the two following operators:

$$x : \psi \mapsto x\psi$$

for the position, and

$$\hbar D_x : \psi \mapsto \frac{\hbar}{i}\nabla\psi$$

for the impulse. Note that these two operators are symmetric with respect to the L^2-scalar product, and are actually self-adjoint with respective domains $\mathcal{F}_\hbar(H^1(\mathbf{R}^3))$ and $H^1(\mathbf{R}^3)$. More generally, any (not necessarily bounded) self-adjoint operator on $L^2(\mathbf{R}^3)$ is called a *quantum observable.*

The situation is the following: We have been able to associate with the classical observable x the quantum observable $\psi \mapsto x\psi$, and with the classical observable ξ the quantum observable $\hbar D_x$. Now a natural question arises: whether such a correspondence between classical observables and quantum observables can be generalized. Namely, given a classical observable $a(x, \xi)$, is there a natural way to associate with it a quantum observable (which could reasonably be denoted by $a(x, \hbar D_x)$)? This is precisely one of the purposes of the pseudodifferential calculus, which also provides an algebraic correspondence between the space of classical observables endowed with the usual multiplication and the space of quantum observables endowed with the composition of operators.

Of course, there are cases where this correspondence can be demonstrated in a very simple way: If $a = a(x)$ does not depend on ξ, then the natural associated quantum observable is just the multiplication by the function a: $\psi \mapsto a\psi$. Similarly, using the fact that $\hbar D_x = \mathcal{F}_\hbar^{-1}\xi\mathcal{F}_\hbar$ (where ξ denotes the operator of multiplication by ξ), in the case where $a = a(\xi)$ does not depend on x, the associated quantum observable is $a(\hbar D_x) := \mathcal{F}_\hbar^{-1}a(\xi)\mathcal{F}_\hbar$. Note that this last definition is consistent with the usual one when $a(\xi)$ is a polynomial in ξ (in which case one obtains a differential operator).

Examples

1. The *kinetic energy* $\frac{1}{2m}\xi^2$ gives rise to the quantum observable $-\frac{\hbar^2}{2m}\Delta$ where $\Delta = \sum_{j=1}^{3} \partial_{x_j}^2$ is the Laplace operator;

2. The total energy $\frac{1}{2m}\xi^2 + V(x)$ is associated with $H = -\frac{\hbar^2}{2m}\Delta + V(x)$,

which is the celebrated *Schrödinger operator.*

Since the physical phenomena essentially consist in exchanges of energy, the study of the operator H is of particular interest in quantum mechanics. The possible energies of a quantum particle submitted to the electric potential V are by definition the eigenvalues of H considered as an operator acting on $L^2(\mathbf{R}^3)$.

Quantum Evolution

In classical mechanics, Newton's law (1.1.1) permits us to predict the evolution of a particle once we know its initial position and momentum. The quantum counterpart can be derived by again using the analogy with optics. From (1.2.2) we get

$$i\frac{\partial \varphi}{\partial t} = \omega\varphi,$$

and by the so-called Planck–Einstein formula (initially obtained experimentally for light, but then generalized to matter by De Broglie in his theory of matter waves, see, e.g., [Mes]), the energy of the wave is given by

$$E = h\nu = \hbar\omega.$$

As a consequence, φ satisfies

$$i\hbar\frac{\partial \varphi}{\partial t} = E\varphi. \tag{1.2.8}$$

Since in quantum mechanics the energy is represented by the operator H, it becomes natural in view of (1.2.8) to require that the evolution of a quantum state be given by the equation

$$i\hbar\frac{\partial \psi}{\partial t} = H\varphi, \tag{1.2.9}$$

which is called the *Schrödinger equation.*

Until now we have dealt with a single quantum particle only. In the case where several (say N) particles are involved in the system that one wants to study, all the previous considerations can be easily generalized by instead denoting by $x = (x^1, \ldots, x^N) \in \mathbf{R}^{3N}$ the set of all the particles, and taking into account that the potential $V(x)$ must be the sum of the external electric field plus all the interactions between the various particles.

1.3 Semiclassical Analysis

As one can guess, the mathematical study of the Schrödinger operator H can be very difficult in general, depending on the potential V that is, involved. As a consequence, one would like to dispose of any kind of approximation physically reasonable, allowing us to predict (at least qualitatively) some of the quantum properties of a system. Moreover, since classical mechanics (which in many aspects is much simpler than quantum mechanics) describes very well most of the common elementary physical phenomena, it is reasonable to hope that quantum mechanics is a kind of generalization of classical mechanics, in the sense that one should be able to recover the classical properties of a system by making some approximation of its quantum properties.

A general principle exists that permits us to give an answer to the two previous questions: The so-called *Bohr correspondence principle* asserts that classical mechanics is nothing but the limit as h tends to 0 of quantum mechanics.

Although this statement remains rather vague, it appears that in many instances it can be both specified and verified. The mathematical branch in which this is performed is commonly called *semiclassical analysis*, and its task consists principally in studying the spectral properties of H asymptotically as $h \to 0$ (that is, for $h > 0$ small enough, without worrying about its actual physical value). However, due to its asymptotic character, semiclassical analysis also allows us to prove mathematically some typical quantum properties, which would be much more difficult to show by taking h fixed (and which are annihilated when the limit $h \to 0$ is taken).

An essential tool of semiclassical analysis is the use of the so-called *pseudodifferential calculus*, which we develop in the next sections. Actually, it may be applied in many other fields, such as the study of the singularities of solutions of partial differential equations (for which it was initially developed; see, e.g., [AlGe, Be1, ChPi, Ho2, KoNi, Sj1, Tr]); the Born–Oppenheimer approximation (used to study the quantum properties of molecules with heavy nuclei, [KMSW]); adiabatic theory (which studies slowly varying systems, [Ju, Mar2]); solid state physics ([GRT, GMS]); scattering theory ([En, SiSo]); and many other aspects of spectral theory ([CdV1, CdV2, DiSj, Du2, Fo, GrSj, He1, He2, Iv, Mas, Ro, Sh, Tay]). Let us stress that in most of these fields h *is not* the Planck constant, but may represent different physical quantities such as the inverse of the square root of the nuclear mass (in the Born–Oppenheimer approximation), the adiabatic parameter (in adiabatic theory), the magnetic field strength (in solid-state physics), the inverse of the square root of the energy (in high-energy spectral problems), or even the inverse of the norm of the position (in scattering theory).

Finally, let us point out that semiclassics also leads to numerous geometric applications that have given rise to a wide area of recent research. Since this aspect will not be considered here, we refer the interesting reader to [Bi, HeSj6, Wi].

1.4 About History

The history of microlocal analysis goes back more than forty years, and since a large number of mathematicians have contributed to the subject, it is quite difficult to describe with precision all the lines of its development. However, the interested reader may find a rich source of information in the historical notes that follow each chapter of the series of books [Ho2] by L. Hörmander. Here we mention that most of the motivation of our notes comes from techniques developed by J. Sjöstrand in [Sj1], and one of our purposes is to give a simplified and unified presentation of them. This has been made possible due to our observation that in a global context, the exponential microlocal estimates that we introduce in Chapter 3 (and which are quite elementary to prove) permit us to get rid of the rather heavy and difficult *pseudodifferential calculus in the complex domain* introduced in [Sj1]. We believe that the main ideas of the proofs in [Sj1] thereby appear in a more enlightened and clearer context, which should allow a better appreciation and understanding of them.

Chapter 2

Semiclassical Pseudodifferential Calculus

2.1 Motivation and Notation

As we have already explained, one of the main motivations of the pseudodifferential calculus is to get an algebraic correspondence between the classical observables and the quantum observables (one calls it a *quantization* of the classical observables). In particular, this would permit us to localize (within the limits allowed by the uncertainty principle) both in position and momentum variables any quantum state ψ: Take a smooth cutoff function $\chi = \chi(x, \xi)$ (that is, $\chi \in C_0^\infty(\mathbf{R}^{2n})$, the space of smooth compactly supported functions on $\mathbf{R}^{2n}$, and χ is close to the characteristic function of some compact subset of $\mathbf{R}^{2n}$). Then its associated quantum observable $\chi(x, \hbar D_x)$ applied to ψ will have the effect of (essentially) cutting off the Cartesian product $\mathrm{Supp}\,\psi \times \mathrm{Supp}\,\widehat{\psi}$ outside $\mathrm{Supp}\,\chi$ (here Supp stands for the support).

Another important feature of this calculus will consist in inverting the so-called *elliptic operators*: If $a(x, \xi)$ is a classical observable that never vanishes (and therefore is invertible in the multiplicative algebra of smooth functions), one would like to be able to invert also its quantization $a(x, \hbar D_x)$. This procedure (called the *construction of a parametrix*) will be possible when a satisfies a little bit more, namely, that it is an invertible element of a special kind of subalgebra of $C^\infty(\mathbf{R}^n)$, called *spaces of symbols* (see the next section).

Many other properties are satisfied by the pseudodifferential operators, and we shall certainly not be exhaustive in these notes, our purpose being to make clear how the things work and to show examples where they can be used. However, there is another application that we want only to mention here, and that is, at the center of a whole field of interest in semiclassical analysis around the so-called *Weyl formula*: The pseudodifferential calculus make it possible to approximate some spectral projectors associated with the Schrödinger operator. We refer to the excellent book of Shubin [Sh] for a detailed approach to this problem.

Now let us fix some standard notation that will be used throughout this book.

If $x = (x_1, \ldots, x_n)$ denotes the current point of $\mathbf{R}^n$, we set

$$\begin{aligned} D_x &= \frac{1}{i}\frac{\partial}{\partial x} = \frac{1}{i}\nabla_x = \frac{1}{i}\partial_x = \frac{1}{i}\left(\frac{\partial}{\partial x_1}, \ldots, \frac{\partial}{\partial x_n}\right), \\ x^2 &= x_1^2 + \ldots + x_n^2; \quad |x| = \sqrt{x^2}, \end{aligned}$$

and, for $\alpha = (\alpha_1, \ldots, \alpha_n) \in \mathbf{N}^n$,

$$\begin{aligned} |\alpha| &= \alpha_1 + \ldots + \alpha_n, \\ \alpha! &= (\alpha_1!) \ldots (\alpha_n!), \\ x^\alpha &= x_1^{\alpha_1} \ldots x_n^{\alpha_n}, \\ D_x^\alpha &= D_{x_1}^{\alpha_1} \ldots D_{x_n}^{\alpha_n}, \\ \partial_x^\alpha &= \partial_{x_1}^{\alpha_1} \ldots \partial_{x_n}^{\alpha_n} \end{aligned}$$

(although the uses of $|x|$ and $|\alpha|$ are not consistent, since $\mathbf{N}^n \subset \mathbf{R}^n$, we use them because they are very traditional, and actually their meaning will always be clear from the context). We also denote by $xy = x \cdot y = \langle x, y \rangle := x_1 y_1 + \ldots + x_n y_n$ the standard scalar product of the two vectors $x = (x_1, \ldots, x_n)$ and $y = (y_1, \ldots, y_n)$ of $\mathbf{R}^n$. Finally, we recall the very useful multidimensional Leibniz formula, valid for any C^∞ functions f, g on $\mathbf{R}^n$ and for any $\alpha \in \mathbf{N}^n$:

$$\partial_x^\alpha (fg) = \sum_{\beta \leq \alpha} \frac{\alpha!}{\beta!(\alpha - \beta)!} \left(\partial_x^\beta f\right) \left(\partial_x^{\alpha-\beta} g\right), \tag{2.1.1}$$

where $\beta \leq \alpha$ means by definition that $\beta_j \leq \alpha_j$ for all $j \in \{1, \ldots, n\}$.

2.2 Spaces of Symbols

Roughly speaking, what we call *symbols* from now on are what we have called *classical observables* until now. In these notes we concentrate on some restricted classes of symbols, in the double sense that they will always be globally defined, and that they will satisfy estimates of a special kind. More general estimates could also be treated, as well as locally defined symbols, but the fact is that most of the difficulties of the theory are already encountered with our simpler classes, so that the reader will in any case learn all the basic techniques of microlocal analysis, without having to care about additional problems that could have the effect of obscuring the concepts (which in some sense remain always relatively simple below the technical discussion). For a very general presentation of the theory, one may consult the very complete series of books of Hörmander [Ho2].

Let $g \in C^\infty(\mathbf{R}^d; \mathbf{R}_+^*)$ (the space of C^∞ functions on $\mathbf{R}^d$ with values in $\mathbf{R}_+^* = (0, \infty)$) satisfying

$$\partial_x^\alpha g = \mathcal{O}(g) \tag{2.2.1}$$

for any $\alpha \in \mathbf{N}^d$ and uniformly on $\mathbf{R}^d$. Such a function is called an *order function* on $\mathbf{R}^d$, and the simplest examples are given by

$$\langle x' \rangle^m := \left(1 + |x'|^2\right)^{m/2},$$

where $m \in \mathbf{R}$ is fixed and $x' = (x_1, \ldots, x_k)$ with $k \leq d$. Other examples are functions such as $e^{\alpha\langle x \rangle}$ with $\alpha \in \mathbf{R}$, or more generally $e^{f(x)}$, where f is smooth and bounded together with all its derivatives. However, a function such as e^{x^2} is not an order function, and neither is any function greater than it (see Exercise 1 at the end of this chapter).

Note that although we have denoted by $x \in \mathbf{R}^d$ the variable of g, in practice we shall almost always have $d = 2n$ and x replaced by (x, ξ), representing the position-momentum variables. In fact, we shall sometimes also deal with $d = 3n$ and x replaced by (x, y, ξ), where the extra variable y plays the role of an integrated variable, in a similar way as when one expresses the distribution kernel of an operator as a function of (x, y).

A first property of this notion of order functions is the following:

Proposition 2.2.1 *If g is an order function on $\mathbf{R}^d$, then so is the function $1/g$.*

Proof This is an easy consequence of the Leibniz formula (2.1.1). Indeed, one has to show that for any $\alpha \in \mathbf{N}^d$, $\partial^\alpha(1/g) = \mathcal{O}(1/g)$. Setting $\tilde{g} = 1/g$ and using the Leibniz formula to differentiate the identity $\tilde{g}g = 1$ α times, the required estimate is easily obtained by induction on $|\alpha|$. ◇

For such an order function g we define the semiclassical space of symbols $S_d(g)$:

Definition 2.2.2 *A function $a = a(x;h)$ defined on $\mathbf{R}^d \times (0, h_0]$ for some $h_0 > 0$ is said to be in $S_d(g)$ if a depends smoothly on x and for any $\alpha \in \mathbf{N}^d$, one has*

$$\partial_x^\alpha a(x;h) = \mathcal{O}(g(x)) \tag{2.2.2}$$

uniformly with respect to $(x,h) \in \mathbf{R}^d \times (0, h_0]$.

In particular, $S_d(1)$ is the set of families of C^∞ functions on $\mathbf{R}^d$ parametrized by some $h \in (0, h_0]$ that are uniformly bounded together with all their derivatives.

Examples

- Any $\chi \in C_0^\infty(\mathbf{R}^d)$ (the space of compactly supported C^∞ functions on $\mathbf{R}^d$) is in $S_d(1)$;
- If $V = V(x) \in S_n(1)$, then the function $\xi^2 + V(x)$ is in $S_{2n}(\langle\xi\rangle^2)$. Note that it corresponds to the total energy (with mass $\frac{1}{2}$) defined in Section 1.2;
- For any $m \in \mathbf{R}$, $\langle x\rangle^m \in S_d(\langle x\rangle^m)$;
- The function $\mathbf{R}^{2n} \ni X = (x,\xi) \mapsto e^{ix\cdot\xi}$ does not belong to $S_{2n}(\langle X\rangle^m)$ for any $m \in \mathbf{R}$, but belongs to any $S_{2n}(e^{\varepsilon\langle X\rangle})$ with $\varepsilon > 0$;
- The functions $e^{-x^2/h}$ and e^{x^2} do not belong to $S_n(g)$ for any order function g on $\mathbf{R}^n$ (just take the value at $x = 0$ of the Laplacian of the first one, and see Exercise 1 of this chapter for the second one).

Note that by Proposition 2.2.1 and the Leibniz formula we have the equivalence

$$a \in S_d(g) \Leftrightarrow \frac{a}{g} \in S_d(1). \tag{2.2.3}$$

We endow $S_d(g)$ with the topology associated with the family of seminorms $N_\alpha(a) = \mathrm{Sup}|\partial^\alpha a|$, and it can be verified easily that this makes $S_d(g)$ a Fréchet space. The basic algebraic properties of the spaces $S_d(g)$ are the following.

Proposition 2.2.3 *Let g_1 and g_2 be two order functions on $\mathbf{R}^d$, and let $a \in S_d(g_1)$, $b \in S_d(g_2)$. Then g_1g_2 is also an order function, and $ab \in S_d(g_1g_2)$.*

Proof This is an obvious consequence of the Leibniz formula. ◇

We define ellipticity as follows:

Definition 2.2.4 *A symbol $a \in S_d(g)$ is said to be* **elliptic** *if there exists a positive constant C_0 such that*

$$|a| \geq \frac{1}{C_0} g$$

uniformly on $\mathbf{R}^d \times (0, h_0]$ (for some $h_0 > 0$).

Then we have the following proposition:

Proposition 2.2.5 *If $a \in S_d(g)$ is elliptic, then $1/a \in S_d(1/g)$.*

Proof Set $b = 1/a$. Then the result is obtained by differentiating iteratively the relation $ab = 1$ and by using the Leibniz formula. ◇

2.3 Semiclassical Expansions of Symbols

In this section we try to specify a little bit more the way in which the symbols may depend on the semiclassical parameter h. First of all, we define the notation $\sim$ (the so-called *asymptotic equivalence* of symbols), which will be used very often in the sequel. Throughout this section g denotes an arbitrary order function on $\mathbf{R}^d$.

Definition 2.3.1 *Let $a \in S_d(g)$ and let $(a_j)_{j \in \mathbf{N}}$ be a sequence of symbols of $S_d(g)$. Then we say that a is* **asymptotically equivalent** *to the formal sum $\sum_{j=0}^{\infty} h^j a_j$ in $S_d(g)$, and we write*

$$a \sim \sum_{j=0}^{\infty} h^j a_j$$

if for any $N \in \mathbf{N}$ and for any $\alpha \in \mathbf{N}^d$ there exist $h_{N,\alpha} > 0$ and $C_{N,\alpha} > 0$ such that

$$\left| \partial^\alpha \left(a - \sum_{j=0}^{N} h^j a_j \right) \right| \leq C_{N,\alpha} h^N g$$

uniformly on $\mathbf{R}^d \times (0, h_{N,\alpha}]$.

In other words, for any $N > 0$ the symbol a can be approximated by $\sum_{j=0}^{N} h^j a_j$ up to a symbol that vanishes together with all its derivatives as h^N when h goes to zero. In practice, the existence of $h_{N,\alpha}$ will not be explicitly written, being referred to as "for h small enough" at the end of an estimate. It will then be implicit that the estimate is valid for h in an interval of the form $(0, h_0]$, where h_0 depends of all the fixed parameters.

In the particular case where all the a_j's are identically zero, we write

$$a = \mathcal{O}(h^\infty) \text{ in } S_d(g) \quad \text{if} \quad a \sim 0 \text{ in } S_d(g).$$

An important and surprising feature is that although a series of the type $\sum_{j=0}^{\infty} h^j a_j$ has no reason to be convergent, one can always find a symbol that is, asymptotically equivalent to it:

Proposition 2.3.2 *Let $(a_j)_{j\in\mathbf{N}}$ be an arbitrary sequence of symbols of $S_d(g)$. Then there exists $a \in S_d(g)$ such that $a \sim \sum_{j=0}^{\infty} h^j a_j$ in $S_d(g)$. Moreover, a is unique up to $\mathcal{O}(h^\infty)$, in the sense that the difference of two such symbols is $\mathcal{O}(h^\infty)$ in $S_d(g)$. Such a symbol a is called a* **resummation** *of the* **formal symbol** $\sum_{j\geq 0} h^j a_j$.

Proof First of all, dividing everything by g and using (2.2.3), we can assume without loss of generality that $g = 1$.

Since the unicity up to $\mathcal{O}(h^\infty)$ is obvious, we concentrate on the existence of a. Then let $\chi \in C_0^\infty(\mathbf{R})$ be such that $\mathrm{Supp}\chi \subset [-2, 2]$, $\chi = 1$ on $[-1, 1]$. We have the following lemma:

Lemma 2.3.3 *There exists a decreasing sequence of positive numbers $(\varepsilon_j)_{j\in\mathbf{N}}$ converging to zero, such that for any $j \in \mathbf{N}$ and $\alpha \in \mathbf{N}^d$ with $|\alpha| \leq j$, one has*

$$\sup_{x\in\mathbf{R}^d} \left| \left(1 - \chi\left(\frac{\varepsilon_j}{h}\right)\right) \partial^\alpha a_j(x; h) \right| \leq h^{-1}$$

for h small enough.

Proof Setting

$$C_j = \sup_{|\alpha|\leq j} \sup_{x\in\mathbf{R}^d} |\partial^\alpha a_j(x; h)|$$

and using the fact that $\left(1-\chi\left(\frac{\varepsilon_j}{h}\right)\right)$ is non zero only for $h \leq \varepsilon_j$, we have

$$h \sup_{x\in\mathbf{R}^d} \left|\left(1-\chi\left(\frac{\varepsilon_j}{h}\right)\right)\partial^\alpha a_j(x;h)\right| \leq C_j\varepsilon_j \leq 1$$

if one has chosen the decreasing sequence $(\varepsilon_j)_{j\geq 0}$ in such a way that $\varepsilon_j \leq \dfrac{1}{C_j}$ for all $j \geq 0$ (one can take, e.g., $\varepsilon_j = \min\{(k+C_k)^{-1}\ ;\ k \leq j\}$). ◇

We then set

$$a(x;h) = \sum_{j\geq 0} h^j \left(1-\chi\left(\frac{\varepsilon_j}{h}\right)\right) a_j(x;h),$$

where actually, the sum contains only a finite number (depending on $h > 0$ fixed) of nonzero terms (since $\varepsilon_j < h$ if j becomes too large). Thus a is a smooth function of $x \in \mathbf{R}^d$, and for any $\alpha \in \mathbf{N}^d$ one has

$$|\partial^\alpha a(x;h)| \leq \sum_{j\leq|\alpha|} h^j|\partial^\alpha a_j(x;h)| + \sum_{j>|\alpha|} h^j \left|\left(1-\chi\left(\frac{\varepsilon_j}{h}\right)\right)\partial^\alpha a_j(x;h)\right|$$

and therefore, using Lemma 2.3.3,

$$|\partial^\alpha a(x;h)| \leq C_\alpha + \sum_{j>|\alpha|} h^{j-1} \leq C'_\alpha,$$

where C_α and C'_α are positive constants.

Thus $a \in S_d(1)$, and for any $\alpha \in \mathbf{N}^d$ and $N \geq |\alpha|$ one has

$$\left|\partial^\alpha\left(a - \sum_{j=0}^N h^j a_j\right)\right| \leq \sum_{j=0}^N h^j \left|\chi\left(\frac{\varepsilon_j}{h}\right)\partial^\alpha a_j\right| + \sum_{j\geq N+1} h^j \left|\left(1-\chi\left(\frac{\varepsilon_j}{h}\right)\right)\partial^\alpha a_j\right|.$$

Using again Lemma 2.3.3. we get

$$\left|\partial^\alpha\left(a - \sum_{j=0}^N h^j a_j\right)\right| \leq \sum_{j=0}^N h^{N+j}\varepsilon_j^{-N} \left|\left(\frac{\varepsilon_j}{h}\right)^N \chi\left(\frac{\varepsilon_j}{h}\right)\right| C_{j,\alpha} + \sum_{j\geq N+1} h^{j-1},$$

where the $C_{j,\alpha}$'s are positive constants. Since the function $\mathbf{R} \ni t \mapsto t^N\chi(t)$ is bounded, we deduce easily from the estimate above that there exists a constant C_N such that for any $h > 0$ sufficiently small,

$$\left|\partial^\alpha\left(a - \sum_{j=0}^N h^j a_j\right)\right| \leq C_N h^N,$$

which completes the proof of Proposition 2.3.2. ◇

Remark 2.3.4 One can generalize the previous notion of equivalence by replacing h^j everywhere it appears by h^{m_j}, where $m_j \in \mathbf{R}$ satisfies $m_j \to +\infty$ as $j \to +\infty$. Then one can prove in the same way an analogous result of resummation.

Remark 2.3.5 In the previous proof we never used the fact that the a_j's are globally defined on $\mathbf{R}^d$. Indeed, a corresponding result for locally defined symbols is easy to state and to verify.

Application: WKB Solutions for the One-Dimensional Schrödinger Operator

Let $V \in C^\infty(\mathbf{R}\,;\mathbf{R})$ and $E \in \mathbf{R}$, and let $x_0 \in \mathbf{R}$ be such that $V(x_0) < E$. Then for x close enough to x_0, one can consider the two smooth functions

$$\varphi_\pm(x) = \pm \int_{x_0}^{x} \sqrt{E - V(y)}dy,$$

which both satisfy $(\varphi_\pm')^2 = E - V$ (the so-called *eikonal equation*). In particular, $\varphi_\pm' \neq 0$ near x_0, and thus $\sqrt{|\varphi_\pm'|}$ is smooth there. Then for any $a = a(x)$ smooth near x_0, one has

$$\begin{aligned}
&\left(-h^2\frac{d^2}{dx^2} + V - E\right)\left(ae^{i\varphi_\pm/h}\right) \\
&= \left[-2iha'\varphi_\pm' - iha\varphi_\pm'' - h^2a'' + \left((\varphi_\pm')^2 + V - E\right)a\right]e^{i\varphi_\pm/h} \\
&= -ih\left[2a'\varphi_\pm' + a\varphi_\pm'' - iha''\right]e^{i\varphi_\pm/h} \\
&= -2ih\sqrt{\varphi_\pm'}\left[\left(a\sqrt{\varphi_\pm'}\right)' - ih\frac{a''}{2\sqrt{\varphi_\pm'}}\right]e^{i\varphi_\pm/h},
\end{aligned}$$

where $\sqrt{\ }$ denotes any determination of the square root on $\mathbf{R}_\pm^*$. Now define recursively a sequence of functions $(a_j^\pm)_{j\geq 0}$ by solving the following differential equations (called the *transport equations*):

$$\left\{\begin{array}{ll}
\left(a_0^\pm\sqrt{\varphi_\pm'}\right)' = 0, & \\
\left(a_j^\pm\sqrt{\varphi_\pm'}\right)' - i\dfrac{(a_{j-1}^\pm)''}{2\sqrt{\varphi_\pm'}} = 0 & \quad (j \geq 1),
\end{array}\right.$$

and denote by $a_\pm(x\,;\,h)$ a resummation of the formal series $\sum_{j\geq 0} h^j a_j^\pm$. Since by construction the formal series $\sum_{j\geq 0} h^j \left[\left(a_j^\pm\sqrt{\varphi'_\pm}\right)' - ih\frac{(a_j^\pm)''}{2\sqrt{\varphi'_\pm}}\right]$ vanishes identically, we see that the function $u_\pm = a_\pm e^{i\varphi_\pm/h}$ satisfies

$$\left(-h^2\frac{d^2}{dx^2} + V(x) - E\right) u_\pm(x\,;\,h) = \mathcal{O}(h^\infty)$$

uniformly for x close enough to x_0 and $h > 0$ small enough.

Such approximate solutions are called *WKB solutions*, from the names of their creators: Wentzel, Kramers, and Brillouin (to whom is sometimes also added Jeffreys).

In particular, taking, for instance,

$$a_0^\pm(x) = \frac{\sqrt{\varphi'_\pm(x_0)}}{\sqrt{\varphi'_\pm(x)}} = \frac{(E - V(x_0))^{1/4}}{(E - V(x))^{1/4}}$$

and, for all $j \geq 1$,

$$a_j^\pm(x) = \frac{i}{\sqrt{\varphi'_\pm(x)}}\int_{x_0}^x \frac{(a_{j-1}^\pm)''(y)}{\sqrt{\varphi'_\pm(y)}}dy = \frac{\pm i}{(E - V(x))^{1/4}}\int_{x_0}^x \frac{(a_{j-1}^\pm)''(y)}{(E - V(y))^{1/4}}dy,$$

one gets an approximate solution $u_\pm$, which in addition satisfies

$$u_\pm(x_0\,;\,h) = 1.$$

2.4 Oscillatory Integrals

In this section we describe the main tool that will be used to define the pseudodifferential operators.

We start by recalling some very basic facts concerning the Fourier transform.

Let $\mathcal{S}(\mathbf{R}^n)$ denote the Schwartz space of smooth functions on $\mathbf{R}^n$ that are rapidly decreasing at infinity together with all their derivatives. For $u \in \mathcal{S}(\mathbf{R}^n)$ we define the (semiclassical) Fourier transform of u as the function

$$\mathcal{F}_h u(\xi) = \widehat{u}(\xi) = \frac{1}{(2\pi h)^{n/2}}\int_{\mathbf{R}^n} e^{-ix\xi/h}u(x)dx, \tag{2.4.1}$$

where $\xi \in \mathbf{R}^n$ and $x\xi$ stands for the scalar product of x and ξ. Then $\mathcal{F}_h$ is an isomorphism on $\mathcal{S}(\mathbf{R}^n)$, the inverse of which is given by

$$\mathcal{F}_h^{-1}v(x) = \frac{1}{(2\pi h)^{n/2}} \int_{\mathbf{R}^n} e^{ix\xi/h} v(\xi) d\xi. \tag{2.4.2}$$

Moreover, $\mathcal{F}_h$ can be extended by duality to an isomorphism of $\mathcal{S}'(\mathbf{R}^n)$ (the Schwartz space of tempered distributions on $\mathbf{R}^n$), which maps $L^2(\mathbf{R}^n)$ into itself, and actually defines an isometry on $L^2(\mathbf{R}^n)$:

$$\forall\, u \in L^2(\mathbf{R}^n)\,, \qquad \|\widehat{u}\|_{L^2(\mathbf{R}^n)} = \|u\|_{L^2(\mathbf{R}^n)}. \tag{2.4.3}$$

One also has

$$\begin{aligned} \mathcal{F}_h(hD_x u) &= \xi \mathcal{F}_h u, \\ \mathcal{F}_h(xu) &= -hD_\xi \mathcal{F}_h u. \end{aligned}$$

Now, if for $u \in \mathcal{S}(\mathbf{R}^n)$ we write $u = \mathcal{F}_h^{-1}\mathcal{F}_h u$, we obtain the formula (where $\delta_{y=x}$ denotes the Dirac measure on $y = x$)

$$\langle \delta_{y=x}, u\rangle = u(x) = \frac{1}{(2\pi h)^n} \int \left(\int e^{i(x-y)\xi/h} u(y) dy \right) d\xi,$$

so that forgetting one moment all the conditions necessary for applying Fubini's theorem, and applying it anyway without worrying about the meaning of what we write, we formally get the strange-looking identity

$$\delta_{y=x} = \frac{1}{(2\pi h)^n} \int e^{i(x-y)\xi/h} d\xi.$$

It is precisely to this kind of divergent integral that we now plan to give a well-defined meaning, and in such a way that all the usual operations valid for the absolutely convergent integrals extend to them, too.

More precisely, let $m \in \mathbf{R}$ and $a = a(x, y, \xi) \in S_{3n}(\langle \xi \rangle^m)$ in the sense of Definition 2.2.2 (where we have taken the particular case $g(x, y, \xi) = \langle \xi \rangle^m$). We are interested in giving a sense to the possibly divergent integral

$$I(a) = \int e^{i(x-y)\xi/h} a(x, y, \xi) d\xi.$$

First of all, we notice that if $m < -n$, then this integral is absolutely convergent and is therefore well-defined. To define it when $m \geq -n$, we are going to interpret it as the *distribution kernel* of an operator, recalling the following basic result (see [Schw2] or, e.g., [Ho2] vol. I).

Theorem 2.4.1 (The Schwartz Kernel Theorem)
If $A : C_0^\infty(\mathbf{R}^n) \to \mathcal{D}'(\mathbf{R}^n)$ is linear and continuous, then there exists a unique $K \in \mathcal{D}'(\mathbf{R}^n \times \mathbf{R}^n)$ such that for any $u, v \in C_0^\infty(\mathbf{R}^n)$ one has

$$\langle Au, v\rangle_{\mathcal{D}',\mathcal{D}} = \langle K, v \otimes u\rangle_{\mathcal{D}',\mathcal{D}}.$$

K is called the **distribution kernel** *of the operator A.*

As usual, we have denoted by $\mathcal{D}'(\mathbf{R}^n)$ the space of distributions on $\mathbf{R}^n$, $v \otimes u \in C_0^\infty(\mathbf{R}^{2n})$ is the tensor product between v and u, and $\langle\ .\ ,\ .\ \rangle_{\mathcal{D}',\mathcal{D}}$ stands for the duality bracket between $\mathcal{D}'$ and $\mathcal{D} = C_0^\infty$.

Now, for $u \in C_0^\infty(\mathbf{R}^n)$ and $a \in S_{3n}(\langle\xi\rangle^m)$ with $m < -n$, we set

$$A_a u(x,h) = \int e^{i(x-y)\xi/h} a(x,y,\xi) u(y)\, dy d\xi$$

and we notice that

$$(1 - h\xi \cdot D_y)\left(e^{i(x-y)\xi/h}\right) = \left(1+\xi^2\right) e^{i(x-y)\xi/h}.$$

Therefore, if we define $L = \dfrac{1}{1+\xi^2}(1 - h\xi D_y) = L(\xi, hD_y)$, we have

$$L\left(e^{i(x-y)\xi/h}\right) = e^{i(x-y)\xi/h}.$$

It is this particular property (which is based on the oscillatory character of the quantity $e^{i(x-y)\xi/h}$ as $|\xi| \to \infty$) that will permit us to give a sense to $I(a)$ for more general a's. In fact, since for any $k \in \mathbf{N}^*$ one has $L^k(e^{i(x-y)\xi/h}) = e^{i(x-y)\xi/h}$, we obtain by making k integrations by parts in the expression of $A_a u(x,h)$,

$$A_a u(x,h) = \int e^{i(x-y)\xi/h} ({}^tL(\xi, hD_y))^k (au)\, dy d\xi =: I_k u(x),$$

where

$$\left({}^tL\right)^k (au) = \left(\frac{1 + h\xi D_y}{1+\xi^2}\right)^k (au) = \mathcal{O}\left(\langle\xi\rangle^{m-k}\right) \tag{2.4.4}$$

uniformly as $|\xi| \to +\infty$. As a consequence, the integral $I_k u(x)$ is absolutely convergent as long as $m - k < -n$, that is, for $m < k - n$. Since moreover $I_{k+\ell} u = I_k u$ for all $\ell \geq 0$, it is therefore reasonable to make the following definition:

Definition 2.4.2 *For all $m \in \mathbf{R}$, $a \in S_{3n}(\langle\xi\rangle^m)$, and $u \in C_0^\infty(\mathbf{R}^n)$, we define*

$$A_a u(x) = \int e^{i(x-y)\xi/h}({}^tL(\xi, hD_y))^k(au)dyd\xi$$

where k is any nonnegative integer greater than $m+n$.

Then we have the following theorem:

Theorem 2.4.3 *A_a defines a continuous linear operator from $C_0^\infty(\mathbf{R}^n)$ to $C^\infty(\mathbf{R}^n)$.*

So we can define the distribution kernel of A_a:

Definition 2.4.4 *We denote by $I(a) = \int e^{i(x-y)\xi/h}a(x,y,\xi)d\xi \in \mathcal{D}'(\mathbf{R}^n \times \mathbf{R}^n)$ the distribution kernel of A_a.*

Remark 2.4.5 One can show that the application $S_{3n}(\langle\xi\rangle^m) \ni a \mapsto I(a) \in \mathcal{D}'(\mathbf{R}^n \times \mathbf{R}^n)$ is continuous, and that it is the only way to define it so that it coincides with the relevant quantity when $m < -n$.

Remark 2.4.6 Here the parameter h does not play any special role.

Proof of Theorem 2.4.3 For $\ell \in \mathbf{N}^*$ arbitrary, let $k = [m] + n + \ell + 1$. Then for $u \in C_0^\infty(\mathbf{R}^n)$ we have $A_a u(x) = I_k u(x)$, and the Lebesgue dominated convergence theorem shows that this function is of class C^ℓ on $\mathbf{R}^n$. Therefore, $A_a u \in C^\infty(\mathbf{R}^n)$, and it depends obviously in a linear way on u. Now if $K \subset \mathbf{R}^n$ is a fixed compact set, and if $\operatorname{Supp} u \subset K$, since $({}^tL)^k$ is a differential operator of order k with coefficients $\mathcal{O}(\langle\xi\rangle^{-k})$ and $\partial_x^\alpha e^{i(x-y)\xi/h} = \mathcal{O}(\langle\xi\rangle^{|\alpha|})$, we obtain immediately for any compact set K' and any $\alpha \in \mathbf{N}^n$,

$$\begin{aligned}\sup_{x\in K'}|\partial^\alpha A_a u(x)| &\leq \sup_{x\in K'}\left|\partial^\alpha I_{[m]+n+|\alpha|+1}u(x)\right| \\ &\leq C_{K,K',\alpha}\sum_{|\beta|\leq[m]+n+|\alpha|+1}\sup_{y\in K}|\partial^\beta u(y)|,\end{aligned}$$

where the constant $C_{K,K',\alpha}$ depends only on K, K', α (actually, here it does not depend on K', but it would be the case if instead of $S_{3n}(\langle\xi\rangle^m)$ one worked in more general symbol spaces, such as $S_{3n}(\langle\xi\rangle^m\langle x\rangle^{m'})$). This completes the proof. ◇

Example

Take $a = 1$. Then one gets for any $k \geq n+1$,

$$\begin{aligned}
A_a u(x) &= \int e^{i(x-y)\xi/h}({}^tL(\xi, hD_y))^k u(y)\,dy\,d\xi \\
&= \int e^{ix\xi/h}\left\langle ({}^tL(\xi, hD_y))^k u, e^{-iy\xi/h}\right\rangle_{\mathcal{E}',\mathcal{E}} d\xi \\
&= \int e^{ix\xi/h}\left\langle u, L(\xi, hD_y)^k e^{-iy\xi/h}\right\rangle_{\mathcal{E}',\mathcal{E}} d\xi \\
&= \int e^{ix\xi/h}\left\langle u, e^{-iy\xi/h}\right\rangle_{\mathcal{E}',\mathcal{E}} d\xi \\
&= (2\pi h)^n \mathcal{F}_h^{-1}\mathcal{F}_h u(x) = (2\pi h)^n u(x),
\end{aligned}$$

where $\langle .,.\rangle_{\mathcal{E}',\mathcal{E}}$ stands for the duality bracket between $\mathcal{E} = C^\infty(\mathbf{R}^n)$ and its dual $\mathcal{E}'$ (which is the space of compactly supported distributions). As a consequence, we have

$$\int e^{i(x-y)\xi/h} d\xi = (2\pi h)^n \delta_{\{y=x\}} \tag{2.4.5}$$

in the sense of oscillatory integrals.

Generalization

One can generalize the previous notion of oscillatory integrals by taking different kinds of functions in the exponential factor. Namely, we say that a function $\varphi = \varphi(x, \theta) \in C^\infty(\mathbf{R}^n \times \mathbf{R}^{n'})$ is a *phase function* if

$$\begin{aligned}
\operatorname{Im}\varphi &\geq 0, \\
\partial^\alpha \varphi &= \mathcal{O}(1) \quad \text{on} \quad \mathbf{R}^{n+n'} \quad \text{when} \quad |\alpha| \geq 2,
\end{aligned} \tag{2.4.6}$$

and there exist $C > 0$ and $\rho > 0$ such that

$$|\nabla_{x,\theta}\varphi| \geq \frac{1}{C}\langle\theta\rangle^\rho \tag{2.4.7}$$

for $|\theta|$ large enough. Then for such a phase function and for $a = a(x, \theta) \in S_{n+n'}(\langle\theta\rangle^m)$ one can define $I_\varphi(a) = \int e^{i\varphi(x,\theta)/h} a(x,\theta)\,d\theta \in \mathcal{D}'(\mathbf{R}^n)$ by the formula

$$\langle I_\varphi(a), u\rangle = \int e^{i\varphi(x,\theta)/h}({}^tL(x, \theta, hD_x, hD_\theta))^k[a(x,\theta)u(x)]\,dx\,d\theta \tag{2.4.8}$$

where $k \geq \dfrac{m+n'}{\rho}$ and

$$L = \frac{1}{1+|\nabla\varphi|^2}\left(1 + h\frac{\partial\bar{\varphi}}{\partial x}D_x + h\frac{\partial\bar{\varphi}}{\partial\theta}D_\theta\right),$$

which satisfies

$$\begin{cases} L(e^{i\varphi/h}) = e^{i\varphi/h}, \\ ({}^tL)^k(au) = \mathcal{O}\left(\dfrac{(1+|\nabla\varphi|)^k\langle\theta\rangle^m}{(1+|\nabla\varphi|^2)^k}\right) = \mathcal{O}(\langle\theta\rangle^{m-k\rho}). \end{cases}$$

Exercise

Prove that $I(a)$ is the limit in $\mathcal{D}'(\mathbf{R}^{2n})$ of the absolutely convergent integral

$$I_\varepsilon(a) = \int e^{i(x-y)\xi/h-\varepsilon\langle\xi\rangle}a(x,y,\xi)d\xi$$

as $\varepsilon \to 0_+$. In particular, all the basic properties of the absolutely convergent integrals remain valid for the oscillatory integrals.

2.5 Pseudodifferential Operators

In view of Theorem 2.4.3, we are now able to give the definition of the pseudodifferential operators we shall work with:

Definition 2.5.1 *For $a \in S_{3n}(\langle\xi\rangle^m)$ and $u \in C_0^\infty(\mathbf{R}^n)$, we set*

$$\boxed{\mathrm{Op}_h(a)u\ (x;h) = \frac{1}{(2\pi h)^n}\int e^{i(x-y)\xi/h}a(x,y,\xi)u(y)\,dy\,d\xi.}$$

Then $\mathrm{Op}_h(a)u \in C^\infty(\mathbf{R}^n)$, and for any $\nu \in \mathbf{R}$ the operator $h^{-\nu}\mathrm{Op}_h(a) : C_0^\infty(\mathbf{R}^n) \to C^\infty(\mathbf{R}^n)$ is called the **semiclassical pseudodifferential operator of symbol $h^{-\nu}a$**. *In this case, $h^{-\nu}\mathrm{Op}_h(a)$ is said to be of degree m and of order ν.*

Examples

(i) Semiclassical differential operators:

In the particular case where a is of the form

$$a(x,y,\xi) = \sum_{|\alpha|\le m} b_\alpha(x)\xi^\alpha$$

with $b_\alpha \in S_n(1)$, we get

$$\mathrm{Op}_h\left(\sum_{|\alpha|\le m} b_\alpha(x)\xi^\alpha\right) = \sum_{|\alpha|\le m} b_\alpha(x)(hD_x)^\alpha.$$

If instead one has $a(x,y,\xi) = \sum_{|\alpha|\le m} b_\alpha(y)\xi^\alpha$, one obtains

$$\mathrm{Op}_h\left(\sum_{|\alpha|\le m} b_\alpha(y)\xi^\alpha\right) = \sum_{|\alpha|\le m} (hD_x)^\alpha b_\alpha(x).$$

More generally, if $a(x,y,\xi) = \sum_{|\alpha|\le m} b_\alpha(x,y)\xi^\alpha$ with $b_\alpha \in S_{2n}(1)$, then for any $u \in C_0^\infty(\mathbf{R}^n)$

$$\mathrm{Op}_h\left(\sum_{|\alpha|\le m} b_\alpha(x,y)\xi^\alpha\right)u(x) = \sum_{|\alpha|\le m} (hD_x)^\alpha \left(b_\alpha(x',x)u(x)\right)|_{x'=x}.$$

(ii) Usual differential operators: If the b_α's are in $S_n(1)$, the differential operator $P = \sum_{|\alpha|\le m} b_\alpha(x)D_x^\alpha$ can be written

$$P = h^{-m}\sum_{j=0}^{m} h^j \sum_{|\alpha|=m-j} b_\alpha(x)(hD_x)^\alpha,$$

and therefore, P is a semiclassical pseudodifferential operator of order m: Its symbol is given by

$$p(x,\xi\,;\,h) = h^{-m}\sum_{j=0}^{m} h^j p_j(x,\xi),$$

where $p_j(x,\xi) = \sum_{|\alpha|=m-j} b_\alpha(x)\xi^\alpha$.

(iii) Inverse of $1 - h^2\Delta$:

Taking $a(x,y,\xi) = (1+\xi^2)^{-1}$, we get an operator that satisfies

$$(1-h^2\Delta)\circ \mathrm{Op}_h\left(\frac{1}{1+\xi^2}\right) = \mathrm{Op}_h\left(\frac{1}{1+\xi^2}\right)\circ(1-h^2\Delta) = 1 \text{ on } C_0^\infty(\mathbf{R}^n).$$

Remark 2.5.2 : Fourier integral operators In the previous definition, if we replace $e^{i(x-y)\xi/h}$ by $e^{i\varphi(x,y,\xi)/h}$, where φ is a phase function on $\mathbf{R}^{2n}\times\mathbf{R}^n$ (in the sense defined by (2.4.6) and (2.4.7)), then we obtain in the same way (via a formula similar to (2.4.8)) an operator $C_0^\infty(\mathbf{R}^n)\to\mathcal{D}'(\mathbf{R}^n)$. Such operators are called *Fourier integral operators* (for short, FIOs). We shall not consider them for the moment, but we shall see some of them in Sections 3.4 and 5.5. A good reference for such operators is [Du1].

Theorem 2.5.3 *For all $a\in S_{3n}(\langle\xi\rangle^m)$, $\mathrm{Op}_h(a)$ can be extended in a unique way to a linear continuous operator $\mathcal{S}(\mathbf{R}^n)\to\mathcal{S}(\mathbf{R}^n)$.*

1st proof Notice that when $k > m+n$, $I_k u$ keeps a meaning for $u\in\mathcal{S}(\mathbf{R}^n)$, and then $I_k u\in C^\infty(\mathbf{R}^n)$. Moreover, using the Leibniz formula as well as the binomial formula, we get

$$x^\beta\partial_x^\alpha I_k u(x) = \sum_{\substack{\alpha'\le\alpha\\ \beta'\le\beta}} c(\alpha',\beta')\int e^{i(x-y)\xi/h}(x-y)^{\beta'}y^{\beta-\beta'}\xi^{\alpha'}({}^tL)^k(\partial_x^{\alpha-\alpha'}au)dy\,d\xi$$

with $c(\alpha',\beta') = (\frac{i}{h})^{|\alpha'|}\dfrac{\alpha!}{\alpha'!(\alpha-\alpha')!}\dfrac{\beta!}{\beta'!(\beta-\beta')!}$.

Therefore, noticing that $e^{i(x-y)\xi/h}(x-y)^{\beta'} = (hD_\xi)^{\beta'}(e^{i(x-y)\xi/h})$ and making β' integrations by parts with respect to ξ, we obtain

$$x^\beta\partial_x^\alpha I_k u(x) = \sum_{\substack{\alpha'\le\alpha\\ \beta'\le\beta}}\mathcal{O}\left(\int\langle y\rangle^{|\beta-\beta'|}\langle\xi\rangle^{|\alpha'|-k+m}\sum_{|\gamma|\le k}|\partial^\gamma u(y)|dy\,d\xi\right).$$

Choosing $k > m+|\alpha|+n$ and writing $\langle y\rangle^{|\beta-\beta'|} = \langle y\rangle^{-n-1}\langle y\rangle^{|\beta-\beta'|+n+1}$, we obtain

$$x^\beta\partial_x^\alpha I_k u(x) = \mathcal{O}\left(\sup_{y\in\mathbf{R}^n}\sum_{|\gamma|\le k}\langle y\rangle^{|\beta|+n+1}|\partial^\gamma u(y)|\right),$$

which proves that $\mathrm{Op}_h(a)u \in \mathcal{S}(\mathbf{R}^n)$ and $|x^\beta \partial_x^\alpha \mathrm{Op}_h(a)u(x)|$ can be estimated by a finite number of seminorms of u in $\mathcal{S}(\mathbf{R}^n)$. Finally, the unicity of the extension of $\mathrm{Op}_h(a)$ to $\mathcal{S}(\mathbf{R}^n)$ comes from the density of $C_0^\infty(\mathbf{R}^n)$ in $\mathcal{S}(\mathbf{R}^n)$. $\diamond$

2nd proof For any $\alpha, \beta \in \mathbf{N}^n$, writing

$$x^\beta \partial_x^\alpha I_k u(x) = \left(\int_{|x-y|\leq \frac{1}{2}|x|} + \int_{|x-y|\geq \frac{1}{2}|x|} \right) x^\beta \partial_x^\alpha \left[e^{i(x-y)\xi/h} ({}^tL)^k (au) \right] dy\, d\xi \tag{2.5.1}$$

with $L = (1+\xi^2)^{-1}(1-h\xi D_y)$, we see that for $k > m+n+|\alpha|$ the first integral is $\mathcal{O}(1)$, because for any $\gamma > 0$.

$$x^\beta \langle \xi \rangle^{m+|\alpha|-k} \langle y \rangle^{-\gamma} = \mathcal{O}(\langle \xi \rangle^{m+|\alpha|-k} \langle y \rangle^{|\beta|-\gamma})$$

uniformly on $\left\{ |x-y| \leq \frac{1}{2}|x| \right\}$, and is therefore integrable with respect to (y,ξ) on $\mathbf{R}^{2n}$ if $\gamma > |\beta| + n$.

On the other hand, setting

$$L' = \frac{1}{1+|x-y|^2}(1 + h(x-y)D_\xi),$$

we see by integrating by parts with respect to ξ that for any $N \in \mathbf{N}$, the second integral can be rewritten as a sum of terms of the type

$$C_{\alpha',\alpha''} \int_{|x-y|\geq \frac{1}{2}|x|} x^\beta e^{i(x-y)\xi/h} ({}^tL')^N \left[\xi^{\alpha'} \partial_x^{\alpha''} ({}^tL)^k (au) \right] dy\, d\xi$$

(with $\alpha' + \alpha'' = \alpha$ and $C_{\alpha',\alpha''}$ constant) and is therefore $\mathcal{O}(1)$ if we take $N \geq |\beta|$.

As a consequence, $\mathrm{Op}_h(a)u \in \mathcal{S}(\mathbf{R}^n)$, and moreover, the previous considerations actually show that $|x^\beta \partial_x^\alpha \mathrm{Op}_h(a)u(x)|$ can be estimated by a finite number of seminorms of u in $\mathcal{S}(\mathbf{R}^n)$. $\diamond$

Remark 2.5.4 Although the first proof can seem more elementary, the interest of the second one is to point out the important fact that the part of the integral corresponding to $|x-y| \geq |x|/2$ (that is, far enough from the diagonal $\{x = y\}$) can be treated in such a way that it becomes essentially irrelevant. Such a phenomenon will reappear several times in the sequel, most often leading to negligible terms (that is, $\mathcal{O}(h^\infty)$ terms).

Now, by duality one also has the following theorem.

Theorem 2.5.5 *For all* $a \in S_{3n}(\langle\xi\rangle^m)$*,* $\mathrm{Op}_h(a)$ *can be extended in a unique way to a linear continuous operator* $\mathcal{S}'(\mathbf{R}^n) \to \mathcal{S}'(\mathbf{R}^n)$*.*

Proof Set $A = \mathrm{Op}_h(a)$. For $v \in \mathcal{S}(\mathbf{R}^n)$, one can define

$$\begin{aligned} {}^tAv(y) &= \frac{1}{(2\pi h)^n}\int e^{i(x-y)\xi/h}a(x,y,\xi)v(x)\,dx\,d\xi \\ &= \frac{1}{(2\pi h)^n}\int e^{i(y-x)\xi/h}a(x,y,-\xi)v(x)\,dx\,d\xi, \end{aligned}$$

which by Theorem 2.5.3 is in $\mathcal{S}(\mathbf{R}^n)$, since $a(x,y,-\xi) \in S_{3n}(\langle\xi\rangle^m)$. Then for all $u, v \in \mathcal{S}(\mathbf{R}^n)$, one has

$$\left\langle u, {}^tAv\right\rangle_{\mathcal{S}',\mathcal{S}} = \langle Au, v\rangle_{\mathcal{S}',\mathcal{S}}$$

(where $\langle .,.\rangle_{\mathcal{S}',\mathcal{S}}$ stands for the duality bracket between $\mathcal{S}'(\mathbf{R}^n)$ and $\mathcal{S}(\mathbf{R}^n)$) and if $u \in \mathcal{S}'(\mathbf{R}^n)$ and $v \in \mathcal{S}(\mathbf{R}^n)$, we define

$$\langle Au, v\rangle_{\mathcal{S}',\mathcal{S}} = \left\langle u, {}^tAv\right\rangle_{\mathcal{S}',\mathcal{S}}.$$

Then the continuity of ${}^tA$ on $\mathcal{S}(\mathbf{R}^n)$ implies that Au defines an element of $\mathcal{S}'(\mathbf{R}^n)$, and it is obvious that if $(u_j)_{j\in\mathbf{N}}$ is a sequence of $\mathcal{S}'(\mathbf{R}^n)$ that converges to $u \in \mathcal{S}'(\mathbf{R}^n)$, then $Au_j \to Au$. As a consequence, the density of $C_0^\infty(\mathbf{R}^n)$ in $\mathcal{S}'(\mathbf{R}^n)$ shows that this is the only way to extend A continuously on $\mathcal{S}'(\mathbf{R}^n)$. $\diamond$

Remark 2.5.6 If $a \in S_{3n}(\langle\xi\rangle^{-\infty}) := \cap_{m\in\mathbf{R}}S_{3n}(\langle\xi\rangle^m)$, then we see that $\mathrm{Op}_h(a)$ maps $\mathcal{S}'(\mathbf{R}^n)$ into $C^\infty(\mathbf{R}^n)$, and for any $u \in \mathcal{S}'(\mathbf{R}^n)$, one has

$$\mathrm{Op}_h(a)u(x) = \frac{1}{(2\pi h)^n}\left\langle u_y, \int e^{i(x-y)\xi/h}a(x,y,\xi)\,d\xi\right\rangle.$$

(To prove this, notice that it is true when $u \in C_0^\infty(\mathbf{R}^n)$. Then use the density of $C_0^\infty(\mathbf{R}^n)$ in $\mathcal{S}'(\mathbf{R}^n)$.) In this case, $\mathrm{Op}(a)$ is said to be a *regularizing* operator.

Remark 2.5.7 : Formal Adjoint of a Pseudodifferential Operator If for $a \in S_{3n}(\langle\xi\rangle^m)$ we set

$$a^*(x,y,\xi) = \overline{a(y,x,\xi)} \in S_{3n}(\langle\xi\rangle^m).$$

Then the operator $(\mathrm{Op}_h(a))^* := \mathrm{Op}_h(a^*)$ satisfies

$$\langle(\mathrm{Op}_h(a))^*u, v\rangle_{L^2} = \langle u, \mathrm{Op}_h(a)v\rangle_{L^2}$$

for all $u, v \in \mathcal{S}(\mathbf{R}^n)$ (where x $\langle .,.\rangle_{L^2}$ denotes the scalar product in the Hilbert space $L^2(\mathbf{R}^n)$ of complex-valued square-integrable functions on $\mathbf{R}^n$, defined by $\langle f, g\rangle := \int f(x)\overline{g(x)}dx$). For this reason, $(\mathrm{Op}_h(a))^*$ is called the *formal adjoint* of $\mathrm{Op}_h(a)$.

2.6 Composition

Thanks to Theorem 2.5.5, there is no theoretical problem in defining the composition of two pseudodifferential operators as an operator on $\mathcal{S}'(\mathbf{R}^n)$. The problem is only to know whether this composed operator is again itself a pseudodifferential operator.

Let $a \in S_{3n}(\langle\xi\rangle^n)$, $b \in S_{3n}(\langle\xi\rangle^{n'})$, $A = \mathrm{Op}_h(a)$, $B = \mathrm{Op}_h(b)$. Then we have formally

$$\begin{aligned}(A \circ B)u(x) &= \frac{1}{(2\pi h)^n}\int e^{i(x-y)\xi/h}a(x,y,\xi)\,Bu(y)\,dy\,d\xi\\ &= \frac{1}{(2\pi h)^n}\int e^{i(x-y')\eta/h}c_h(x,y',\eta)\,u(y')\,dy'\,d\eta\end{aligned}$$

with

$$c_h(x,y',\eta) = \frac{1}{(2\pi h)^n}\int e^{i(x-y)(\xi-\eta)/h}a(x,y,\xi)\,b(y,y',\eta)\,dy\,d\xi.$$

So, to show that $A \circ B$ is a pseudodifferential operator, it is enough to prove that c_h is in some $S_{3n}(\langle\xi\rangle^m)$. The main tool for that is, the following:

Theorem 2.6.1 (Stationary Phase Theorem) *For $d \in \mathbf{N}^*$, let Q be a $d \times d$ nondegenerate real symmetric matrix. Then for all $u \in C_0^\infty(\mathbf{R}^d)$ and for all $N \geq 1$, one has*

$$\int e^{ix\cdot Qx/2h}u(x)dx = \sum_{k=0}^{N-1}\frac{(2\pi)^{d/2}h^{k+\frac{d}{2}}e^{i\frac{\pi}{4}\operatorname{sgn}Q}}{(2i)^k|\det Q|^{1/2}k!}((D_x\cdot Q^{-1}D_x)^k u)(0) + S_N(u,h)$$

with

$$|S_N(u,h)| \leq \frac{Ch^{N+\frac{d}{2}}}{2^N N!\sqrt{|\det Q|}}\sum_{|\alpha|\leq d+1}\left\|\partial_x^\alpha(D_x\cdot Q^{-1}D_x)^N u\right\|_{L^1(\mathbf{R}^d)},$$

where $C > 0$ depends only on d, and $\operatorname{sgn} Q$ denotes the signature of Q.

Proof Let $\mathcal{F}_1$ denote the d-dimensional Fourier transform with $h = 1$, that is,

$$\mathcal{F}_1 : u \mapsto \frac{1}{(2\pi)^{d/2}}\int e^{-ix\xi}u(x)\,dx.$$

First we have a lemma:

Lemma 2.6.2

$$\mathcal{F}_1^{-1}\left(e^{ix\cdot Qx/2h}\right) = h^{d/2}e^{i\frac{\pi}{4}\operatorname{sgn} Q}|\det Q|^{-1/2}e^{-ih(\xi\cdot Q^{-1}\xi)/2}.$$

Proof of the Lemma If $a \in \mathbf{C}$ satisfies $\operatorname{Re} a > 0$, one has on $\mathbf{R}$

$$\mathcal{F}_1^{-1}\left(e^{-ax^2/2}\right) = \mathcal{F}_1\left(e^{-ax^2/2}\right) = \frac{1}{\sqrt{a}}e^{-\xi^2/2a}.$$

Then, taking $a = \varepsilon \mp i\mu$ with $\mu > 0$, and taking the limit $\varepsilon \to 0$, one gets

$$\mathcal{F}_1^{-1}\left(e^{\pm i\mu x^2/2}\right) = \frac{1}{\sqrt{\mu}}e^{\pm i\pi/4}e^{\pm\xi^2/2i\mu},$$

from which the lemma follows after diagonalization of Q in an orthonormal basis of $\mathbf{R}^d$. ◇

The interest of the previous lemma is to transform the rapid oscillations of $e^{ix\cdot Qx/2h}$ (as $h \to 0$) into the slow ones of $e^{-ih(\xi\cdot Q^{-1}\xi)/2}$. In fact, by definition of the Fourier transform on $\mathcal{S}'(\mathbf{R}^d)$ we have

$$\begin{aligned}\int e^{ix\cdot Qx/2h}u(x)dx &= \left\langle e^{ix\cdot Qx/2h}, u\right\rangle_{S',S} = \left\langle \mathcal{F}_1^{-1}(e^{ix\cdot Qx/2h}), \mathcal{F}_1 u\right\rangle_{S',S} \\ &= h^{d/2}e^{i\frac{\pi}{4}\operatorname{sgn} Q}|\det Q|^{-1/2}\int e^{-ih\xi\cdot Q^{-1}\xi/2}\widehat{u}(\xi)\,d\xi,\end{aligned}$$

and since for all $N \in \mathbf{N}$, $|\partial_t^N(e^{it})| = 1$, we have by Taylor's formula

$$\left|e^{it} - \sum_{k=0}^{N-1}\frac{(it)^k}{k!}\right| \leq \frac{|t|^N}{N!} \qquad \forall t \in \mathbf{R}.$$

In particular,

$$e^{-ih\xi\cdot Q^{-1}\xi/2} = \sum_{k=0}^{N-1}\frac{1}{k!}\left(\frac{h}{2i}(\xi\cdot Q^{-1}\xi)\right)^k + R_N(\xi,h)$$

with

$$|R_N(\xi,h)| \leq \frac{h^N}{2^N N!}\left|(\xi\cdot Q^{-1}\xi)\right|^N.$$

Since moreover,

$$\int(\xi\cdot Q^{-1}\xi)^k\widehat{u}(\xi)\,d\xi = (2\pi)^{n/2}\left((D_x\cdot Q^{-1}D_x)^k u\right)(0),$$

we obtain the expected formula with a remainder term S_N satisfying

$$|S_N| \le \frac{h^{N+d/2}}{2^N N!} |\det Q|^{-1/2} \int |(\xi \cdot Q^{-1}\xi)^N \widehat{u}(\xi)| d\xi.$$

Then, using the well-known inequality

$$\|\mathcal{F}_1 v\|_{L^1(\mathbf{R}^d)} \le \sum_{|\alpha|\le n+1} \|\partial^\alpha v\|_{L^1(\mathbf{R}^d)},$$

which can be obtained, e.g., by noticing that

$$\mathcal{F}_1 v = (2\pi)^{-d/2}(1+\xi^2)^{-d-1} \int e^{-ix\xi}(1-i\xi D_x)^{d+1} v(x) dx,$$

and the fact that

$$(\xi \cdot Q^{-1}\xi)^N \widehat{u}(\xi) = \mathcal{F}_1\left((D_x \cdot Q^{-1} D_x)^N u\right)(\xi)$$

the results follows. ◇

In the sequel, we shall essentially use the following particular case of Theorem 2.6.1.

Corollary 2.6.3 *For all $u \in C_0^\infty(\mathbf{R}^{2n})$ and $N \ge 1$, one has*

$$\frac{1}{(2\pi h)^n} \int e^{-ixy/h} u(x,y)\, dx\, dy = \sum_{k=0}^{N-1} \frac{h^k}{i^k k!} \left(\sum_{j=1}^n \partial_{x_j}\partial_{y_j}\right)^k u(0,0) + S_N$$

with

$$|S_N| \le \frac{Ch^N}{N!} \sum_{|\alpha+\beta|\le 2n+1} \left\|\partial_x^\alpha \partial_y^\beta (\partial_x\partial_y)^N u\right\|_{L^1(\mathbf{R}^{2n})},$$

where $C > 0$ depends only on n.

Proof It is a direct application of Theorem 2.6.1 with $d = 2n$ and $Q = Q^{-1} = \begin{pmatrix} 0 & -I \\ -I & 0 \end{pmatrix}$, which satisfies $\operatorname{sgn} Q = 0$, $|\det Q| = 1$. ◇

Remark 2.6.4 Notice that for all $k \ge 0$ one has

$$\left(\sum_{j=1}^n \partial_{x_j}\partial_{y_j}\right)^k = \sum_{|\alpha|=k} \frac{k!}{\alpha!} \partial_x^\alpha \partial_y^\alpha,$$

so that the expansion found in Corollary 2.6.3 can be rewritten as

$$\sum_{|\alpha|\leq N-1} \frac{h^{|\alpha|}}{i^{|\alpha|}\alpha!}\partial_x^\alpha \partial_y^\alpha u(0,0) + S_N.$$

Now we can prove the main result of this section:

Theorem 2.6.5 (Theorem of Composition) *For all $a \in S(\langle\xi\rangle^m)$ and $b \in S(\langle\xi\rangle^{m'})$ there exists $c \in S(\langle\xi\rangle^{m+m'})$ such that*

$$\mathrm{Op}_h(a) \circ \mathrm{Op}_h(b) = \mathrm{Op}_h(c).$$

Moreover, a possible choice for c is given by the oscillatory integral

$$c(x,y,\xi) = \frac{1}{(2\pi h)^n}\int e^{i(x-z)(\eta-\xi)/h} a(x,z,\eta) b(z,y,\xi)\, dz\, d\eta \stackrel{\text{def}}{=} a\widetilde{\#}b(x,y,\xi),$$

which satisfies

$$a\widetilde{\#}b(x,y,\xi) \sim \sum_{|\alpha|\geq 0} \frac{h^{|\alpha|}}{i^{|\alpha|}\alpha!}\partial_z^\alpha \partial_\eta^\alpha (a(x,z,\eta)b(z,y,\xi))\Big|_{\substack{z=x\\ \eta=\xi}} \text{ in } S_{3n}(\langle\xi\rangle^{m+m'}) .$$

Remark 2.6.6 This composition has a meaning as well on $\mathcal{S}(\mathbf{R}^n)$ or on $\mathcal{S}'(\mathbf{R}^n)$.

Remark 2.6.7 Notice that the asymptotics of $a\widetilde{\#}b$ retain a meaning even if a and b are in the largest class of h-dependent C^∞ functions satisfying

$$|\partial^\alpha a| = \mathcal{O}(h^{-|\alpha|\mu}\langle\xi\rangle^m) \; ; \; |\partial^\alpha b| = \mathcal{O}(h^{-|\alpha|\mu}\langle\xi\rangle^{m'})$$

where $\mu \in [0, \frac{1}{2})$ is fixed. In fact, it can be shown that the theorem remains valid for such classes of symbols, and we refer to the book of M.A. Shubin [Sh] for more details about this, as well as for applications (such as the so-called *semiclassical Weyl formula*).

Remark 2.6.8 As we shall see later on, the choice of c is not unique.

Proof of the Theorem Making integrations by parts and using the same decomposition as in (2.5.1), we see that for $u \in C_0^\infty$ we have

$$\mathrm{Op}_h(b)u(z) = \lim_{\substack{\varepsilon\to 0_+\\ \delta\to 0_+}} {}_{\mathcal{S}(\mathbf{R}^n)} \frac{1}{(2\pi h)^n}\int e^{i(z-y)\xi/h - \varepsilon\langle\xi\rangle - \delta\langle z\rangle} b(z,y,\xi)u(y)\,dy\,d\xi, \quad (2.6.1)$$

where $\lim_{\mathcal{S}(\mathbf{R}^n)}$ means that the limit takes place for the topology of $\mathcal{S}(\mathbf{R}^n)$. As a consequence, the continuity of $\mathrm{Op}(a)$ on $\mathcal{S}(\mathbf{R}^n)$ gives

$$\begin{aligned}&(2\pi h)^{2n}\mathrm{Op}_h(a)\circ\mathrm{Op}_h(b)u(x)\\&\quad=\lim_{\substack{\varepsilon\to 0_+\\ \delta\to 0_+}}\int e^{i(x-z)\eta/h}a(x,z,\eta)\left(\int e^{i(z-y)\xi/h-\varepsilon\langle\xi\rangle-\delta\langle z\rangle}b(z,y,\xi)u(y)dy\,d\xi\right)dz\,d\eta\end{aligned}$$

which by a similar argument can be rewritten as

$$\begin{aligned}&(2\pi h)^{2n}\mathrm{Op}_h(a)\circ\mathrm{Op}_h(b)u(x)\\&\quad=\lim_{\substack{\varepsilon\to 0_+\\ \delta\to 0_+}}\int e^{i(x-z)\eta/h+i(z-y)\xi/h-\varepsilon\langle\xi\rangle-\delta\langle z\rangle-\delta\langle\eta\rangle}a(x,z,\eta)b(z,y,\xi)u(y)dy\,d\xi\,dz\,d\eta,\end{aligned}$$

and therefore

$$\mathrm{Op}_h(a)\circ\mathrm{Op}_h(b)u(x)=\lim_{\varepsilon\to 0_+}\lim_{\delta\to 0_+}\frac{1}{(2\pi h)^n}\int e^{i(x-y)\xi/h-\varepsilon\langle\xi\rangle}c_\delta(x,y,\xi)u(y)dy\,d\xi \tag{2.6.2}$$

with

$$c_\delta(x,y,\xi)=\frac{1}{(2\pi h)^n}\int e^{i(x-z)(\eta-\xi)/h-\delta\langle z\rangle-\delta\langle\eta\rangle}a(x,z,\eta)b(z,y,\xi)dz\,d\eta.$$

As a consequence, by the dominated convergence theorem it is enough to prove that $c_\delta=\mathcal{O}(\langle\xi\rangle^{m+m'})$ uniformly with respect to δ, that for all $(x,y,\xi)\in\mathbf{R}^{3n}$, $c_\delta(x,y,\xi)$ has a limit $c_0(x,y,\xi)$ as $\delta\to 0_+$ (so that the first limit $\delta\to 0_+$ can be taken in (2.6.2), leading to a convergent integral), and that $c_0\in S_{3n}(\langle\xi\rangle^{m+m'})$ (so that the second limit $\varepsilon\to 0_+$ can be taken in (2.6.2), leading to an oscillatory integral). To do so, the idea consists in splitting the integral into three pieces (by means of cutoff functions), in such a way that $|x-z|$ (respectively $|\eta-\xi|$) remains away from zero in the first (respectively second) piece, while both $|x-z|$ and $|\eta-\xi|$ remain bounded in the last piece. Then one shows that the first two parts are essentially negligible, while the stationary phase theorem can be applied to the third one. Now let us give the details.

Set

$$L_1=\left(1+\frac{|\eta-\xi|^2}{h^2}+\frac{|x-z|^2}{h^2}\right)^{-1}\left(1-\frac{(\eta-\xi)}{h}D_z+\frac{(x-z)}{h}D_\eta\right)$$

and let $\chi_1 \in C_0^\infty(\mathbf{R})$, $\chi_1(s) = 1$ for $|s| \leq 1$, $\chi_1(s) = 0$ for $|s| \geq 2$. For $x, y \in \mathbf{R}^n$, we set $\chi(x,y) = \chi_1(|x-y|)$. Then for any $k \geq |m| + 2n + 1$ one has

$$\begin{aligned} c_\delta(x,y,\xi) &= \frac{1}{(2\pi h)^n} \int e^{i(x-z)(\eta-\xi)/h} ({}^tL_1)^k (e^{-\delta\langle z\rangle - \delta\langle\eta\rangle} a(x,z,\eta) b(z,y,\xi)) dz\, d\eta \\ &= d_\delta(x,y,\xi) + e_\delta(x,y,\xi) + f_\delta(x,y,\xi), \end{aligned}$$

where

$$\begin{aligned} &(2\pi h)^n d_\delta(x,y,\xi) \\ &\quad = \int e^{i(x-z)(\eta-\xi)/h} ({}^tL_1)^k \left((1-\chi(\xi,\eta)) e^{-\delta\langle z\rangle - \delta\langle\eta\rangle} a(x,z,\eta) b(z,y,\xi)\right) dz\, d\eta \\ &\quad = \int_{|\eta-\xi|\geq 1} \mathcal{O}\left(\frac{\langle\eta\rangle^m \langle\xi\rangle^{m'}}{(1+h^{-1}|\eta-\xi| + h^{-1}|x-z|)^k}\right) dz\, d\eta \\ &\quad = \int \mathcal{O}\left(\frac{\langle\eta\rangle^m \langle\xi\rangle^{m'}}{(1+\frac{1+|\eta-\xi|}{2h})^{k-n-1/2}}\right) d\eta, \end{aligned}$$

and thus in the case $m \geq 0$,

$$\begin{aligned} (2\pi h)^n d_\delta(x,y,\xi) &= \int \mathcal{O}\left(h^{k-n-1/2} \frac{(\langle\xi\rangle + \langle\eta-\xi\rangle)^m \langle\xi\rangle^{m'}}{\langle\eta-\xi\rangle^{k-n-1/2}}\right) d\eta \\ &= \mathcal{O}\left(h^{k-n-1/2} \langle\xi\rangle^{m+m'}\right). \end{aligned}$$

In the case $m < 0$, one splits the integral into the two regions $\{|\eta| \geq \langle\xi\rangle/2\}$ and $\{|\eta| \leq \langle\xi\rangle/2\}$. In the first region one has $\langle\eta\rangle^m = \mathcal{O}(\langle\xi\rangle^m)$, and therefore one gets the same estimate as before. In the second region one has $\langle\eta-\xi\rangle \geq \langle\xi\rangle/C$ for some positive constant C, and therefore the corresponding integral can be estimated by $\mathcal{O}\left(h^{k-n-1/2} \langle\xi\rangle^{m'-(k-2n-1)}\right)$.

Similarly,

$$\begin{aligned} &(2\pi h)^n e_\delta(x,y,\xi) \\ &\quad = \int e^{i(x-z)(\eta-\xi)/h} ({}^tL_1)^k \Big[\chi(\xi,\eta)(1-\chi(x,z)) e^{-\delta\langle z\rangle - \delta\langle\eta\rangle} \\ &\qquad\qquad \times a(x,z,\eta) b(z,y,\xi)\Big] dz\, d\eta \\ &\quad = \mathcal{O}\left(h^{k-n-1/2} \langle\xi\rangle^{m+m'}\right) \end{aligned}$$

for all $k \geq |m| + 2n + 1$, and uniformly with respect to $(x, y, \xi) \in \mathbf{R}^{3n}$ and $\delta > 0$. Actually, the same argument also gives that for any $\alpha \in \mathbf{N}^{3n}$,

$$|\partial^\alpha d_\delta(x, y, \xi)| + |\partial^\alpha e_\delta(x, y, \xi)| = \mathcal{O}\left(h^\infty \langle \xi \rangle^{m+m'}\right) \tag{2.6.3}$$

uniformly with respect to $(x, y, \xi) \in \mathbf{R}^{3n}$ and $\delta > 0$ (indeed, the effect of differentiating d_δ or e_δ is just to make some new powers of $|x - z|$ or $|\eta - \xi|$ appear in the estimates, and this can be overcome by taking k greater).

So it remains to study the last term f_δ, which, by integrations by parts, can be written as

$$f_\delta(x, y, \xi) = \frac{1}{(2\pi h)^n} \int e^{i(x-z)(\eta-\xi)/h} \chi(\xi, \eta) \chi(x, z) e^{-\delta\langle z \rangle - \delta\langle \eta \rangle} a(x, z, \eta) b(z, y, \xi) dz\, d\eta.$$

Making the change of variables

$$\begin{cases} z' = z - x, \\ \eta' = \eta - \xi, \end{cases}$$

we get

$$f_\delta(x, y, \xi) = \frac{1}{(2\pi h)^n} \int e^{-iz'\eta'/h} u^\delta_{x,y,\xi}(z', \eta') dz' d\eta'$$

with

$$\begin{aligned} u^\delta_{x,y,\xi}(z', \eta') &= \chi(\xi, \eta' + \xi) \chi(x, z' + x) e^{-\delta\langle z' + x \rangle - \delta\langle \eta' + \xi \rangle} \\ &\quad \times a(x, z' + x, \eta' + \xi) b(z' + x, y, \xi) \in C_0^\infty(\mathbf{R}^n_{z'} \times \mathbf{R}^n_{\eta'}). \end{aligned}$$

Then we can apply the stationary phase theorem to this integral (in the version given by Corollary 2.6.3), and we obtain for all $N \geq 1$,

$$f_\delta(x, y, \xi) = \sum_{|\alpha| \leq N-1} \frac{h^{|\alpha|}}{i^{|\alpha|} \alpha!} \partial_z^\alpha \partial_\eta^\alpha u^\delta_{x,y,\xi}(z, \eta) \Big|_{\substack{z=0 \\ \eta=0}} + S_N \tag{2.6.4}$$

with

$$\begin{aligned} |S_N| &\leq \frac{Ch^N}{N!} \sum_{|\alpha+\beta| \leq 2n+1} \left\| \partial_z^\alpha \partial_\eta^\beta (\partial_z \partial_\eta)^N u^\delta_{x,y,\xi} \right\|_{L^1(\mathbf{R}^n_z \times \mathbf{R}^n_\eta)} \\ &= \mathcal{O}\left(h^N \int_{\substack{|\eta - \xi| \leq 2 \\ |x - z| \leq 2}} \langle \eta \rangle^m \langle \xi \rangle^{m'} dz\, d\eta \right) \\ &= \mathcal{O}\left(h^N \langle \xi \rangle^{m+m'} \right) \end{aligned}$$

uniformly. Doing the same procedure for $\partial^\gamma f_\delta$ ($\gamma \in \mathbf{N}^{3n}$ arbitrary), we get, in particular,

$$|\partial^\gamma f_\delta(x,y,\xi)| = \mathcal{O}(\langle\xi\rangle^{m+m'}) \tag{2.6.5}$$

uniformly with respect to $\delta > 0$ and $(x,y,\xi) \in \mathbf{R}^{3n}$.

Moreover, since for $k \geq m+2n+1$ one has

$$\left\|({}^tL_1)^k\left(e^{-\delta\langle z\rangle-\delta\langle\eta\rangle}a(x,z,\eta)b(z,y,\xi)\right)\right\|_{L^1(\mathbf{R}_z^n\times\mathbf{R}_\eta^n)} = \mathcal{O}_{x,y,\xi}(1)$$

uniformly with respect to δ, we get by the dominated convergence theorem:

$$c_\delta(x,y,\xi) \to c_0(x,y,\xi) \text{ as } \delta \to 0_+,$$

where

$$c_0(x,y,\xi) = \frac{1}{(2\pi h)^n}\int e^{i(x-z)(\eta-\xi)/h}({}^tL_1)^k(a(x,z,\eta)b(z,y,\xi))dz\,d\eta.$$

Since the estimates (2.6.3) and (2.6.5) are uniform with respect to δ, we also have

$$c_0 \in S_{3n}(\langle\xi\rangle^{m+m'}),$$

and finally, we deduce from (2.6.2) that

$$\begin{aligned}\mathrm{Op}_h(a)\circ\mathrm{Op}_h(b)u(x) &= \lim_{\varepsilon\to 0_+}\frac{1}{(2\pi h)^n}\int e^{i(x-y)\xi/h-\varepsilon\langle\xi\rangle}c_0(x,y,\xi)u(y)dy\,d\xi\\ &= \frac{1}{(2\pi h)^n}\int e^{i(x-y)\xi/h}c_0(x,y,\xi)u(y)dy\,d\xi,\end{aligned}$$

where the last integral has to be interpreted as an oscillatory one. Taking also the limit $\delta \to 0_+$ into (2.6.4), we obtain the semiclassical asymptotic expansion of $c_0(x,y,\xi)$, and this completes the proof of Theorem 2.6.5. ◇

Remark 2.6.9 In the case where $a(x,y,\xi)$ is polynomial with respect to ξ (that is, when $\mathrm{Op}_h(a)$ is a differential operator), the asymptotic formula giving $a\widetilde{\#}b$ becomes an *exact* formula (and indeed, the sum becomes finite): This can be easily seen from the definition of $c(x,y,\xi)$ or even more directly by computing $(hD_x)^\alpha\mathrm{Op}_h(b)$ for any $\alpha \in \mathbf{N}^n$.

As a first application of this theorem of composition we have the following construction of a so-called *parametrix* for the elliptic pseudodifferential operators.

Proposition 2.6.10 *Let $m \in \mathbf{R}$ and let $a \in S_{3n}(\langle\xi\rangle^m)$ be an elliptic symbol in the sense of Definition 2.2.4. Then there exists $b \in S_{3n}(\langle\xi\rangle^{-m})$ such that*

$$\begin{cases} \mathrm{Op}_h(a) \circ \mathrm{Op}_h(b) = 1 + \mathrm{Op}_h(r), \\ \mathrm{Op}_h(b) \circ \mathrm{Op}_h(a) = 1 + \mathrm{Op}_h(r'), \end{cases}$$

with $r, r' = \mathcal{O}(h^\infty)$ in $S_{3n}(1)$.

Proof By Proposition 2.2.5, we know that $\dfrac{1}{a} \in S_{3n}(\langle\xi\rangle^{-m})$. Then, setting $b_0 = \dfrac{1}{a}$ and using the expansion of $a\tilde{\#}b$ given in the theorem, it is possible to define $b_j \in S_{3n}(\langle\xi\rangle^{-m})$ $(j = 1, 2, \ldots)$ recursively, in such a way that if $b \sim \sum h^j b_j$, then

$$a\tilde{\#}b = 1 + \mathcal{O}(h^\infty) \text{ in } S_{3n}(1).$$

In the same way, one finds $b' \in S_{3n}(\langle\xi\rangle^{-m})$ such that

$$b'\tilde{\#}a = 1 + \mathcal{O}(h^\infty) \text{ in } S_{3n}(1),$$

and by Theorem 2.6.5, this implies

$$\begin{cases} \mathrm{Op}_h(a) \circ \mathrm{Op}_h(b) = 1 + \mathrm{Op}_h(r), \\ \mathrm{Op}_h(b') \circ \mathrm{Op}_h(a) = 1 + \mathrm{Op}_h(r'), \end{cases}$$

with $r, r' = \mathcal{O}(h^\infty)$ in $S_{3n}(1)$. As a consequence, using the associativity of the composition, we get

$$(1 + \mathrm{Op}_h(r'))\mathrm{Op}_h(b) = \mathrm{Op}_h(b')(1 + \mathrm{Op}_h(r)),$$

and therefore, using again Theorem 2.6.5,

$$\mathrm{Op}_h(b) = \mathrm{Op}_h(b') + \mathrm{Op}_h(b')\mathrm{Op}_h(r) - \mathrm{Op}_h(r')\mathrm{Op}_h(b) = \mathrm{Op}_h(b') + \mathrm{Op}_h(r_1)$$

with $r_1 = \mathcal{O}(h^\infty)$ in $S_{3n}(\langle\xi\rangle^{-m})$. In particular,

$$\mathrm{Op}_h(b)\mathrm{Op}_h(a) = \mathrm{Op}_h(b')\mathrm{Op}_h(a) + \mathrm{Op}_h(r_1)\mathrm{Op}_h(a) = 1 + \mathrm{Op}_h(r_2)$$

with $r_2 = \mathcal{O}(h^\infty)$ in $S_{3n}(1)$. Therefore, b solves the problem. ◇

Remark 2.6.11 As we shall see later on, the fact that $r = \mathcal{O}(h^\infty)$ in $S_{3n}(1)$ implies that $\mathrm{Op}_h(r)$ is $\mathcal{O}(h^\infty)$ in the $\mathcal{L}(L^2)$-norm of bounded operators on $L^2(\mathbf{R}^n)$. Then Proposition 2.6.10 permits one to find the inverse of an elliptic operator on $L^2(\mathbf{R}^n)$ in terms of a Neumann series of pseudodifferential operators.

2.7 Change of Quantization - Symbolic Calculus

As we have seen in the introduction, the classical observables are given in physics as functions of the positions and the momenta of a set of particles. In particular, if $x \in \mathbf{R}^n$ denotes the positions, these functions depend on $2n$ variables only. Then it could seem more convenient to work with pseudo-differential operators with symbols of the form $a = a(x, \xi)$ depending on $2n$ variables only. In this section we shall see that it is always possible, and that actually one can construct one to one correspondences between pseudodifferential operators and symbols depending on $2n$ variables. However, due to the noncommutativity of the operators hD_{x_j} and x_j, there is no unique way to quantize such symbols (e.g., the symbol $x_j\xi_j = \xi_j x_j$ could be associated with both operators $x_j \cdot hD_{x_j}$ and $hD_{x_j} \cdot x_j = x_j \cdot hD_{x_j} - ih$).

Noticing that for $a \in S_{2n}(\langle\xi\rangle^m)$ and $t \in [0,1]$ we have $a((1-t)x + ty, \xi) \in S_{3n}(\langle\xi\rangle^m)$, we set

$$\boxed{\mathrm{Op}_h^t(a) := \mathrm{Op}_h(a((1-t)x + ty, \xi)).}$$

The values $t = 0$, $t = \frac{1}{2}$ and $t = 1$ play a particular role, and they are respectively called:

$t = 0$: "standard" quantization, or "left" quantization;
$t = \frac{1}{2}$: Weyl quantization (also denoted by $\mathrm{Op}_h^W(a)$);
$t = 1$: "right" quantization.

The Weyl quantization is particularly useful in quantum mechanics because it has the pleasant property that when a is real-valued, then $\mathrm{Op}^W(a)$ is symmetric with respect to the $L^2(\mathbf{R}^n)$-scalar product.

In practice, to simplify the computations it is always better to try to remain in a given quantization, that is, with a given fixed value of t. This appears to be possible because of the following two theorems:

Theorem 2.7.1 *Let $b = b(x, y, \xi) \in S_{3n}(\langle\xi\rangle^m)$ and $t \in [0,1]$. Then there exists a unique $b_t(x, \xi) \in S_{2n}(\langle\xi\rangle^m)$ such that*

$$\mathrm{Op}_h(b) = \mathrm{Op}_h^t(b_t).$$

Moreover, b_t is given by the oscillatory integral

$$b_t(x,\xi) = \frac{1}{(2\pi h)^n}\int_{\mathbf{R}^{2n}} e^{i(\xi'-\xi)\theta/h} b(x+t\theta, x-(1-t)\theta, \xi')\, d\xi'\, d\theta \tag{2.7.1}$$

and satisfies

$$b_t(x,\xi) \sim \sum_{\alpha\in\mathbf{N}^n} \frac{(-1)^{|\alpha|}h^{|\alpha|}}{i^{|\alpha|}\alpha!} \partial_\xi^\alpha \partial_\theta^\alpha b(x+t\theta, x-(1-t)\theta,\xi)\Big|_{\theta=0} \text{ in } S_{2n}(\langle\xi\rangle^m).$$

Here b_t is called the **symbol of index** *t (or t-***symbol***) of $B = \mathrm{Op}_h(b)$, and it is denoted by*

$$b_t = \sigma_t(B).$$

When $t=\frac{1}{2}$, $b_{\frac{1}{2}} = b^W$ is also-called the Weyl symbol of $\mathrm{Op}(b)$.

Proof We try to find $b_t \in S_{2n}(\langle\xi\rangle^m$ such that

$$\int e^{i(x-y)\xi/h} b(x,y,\xi)\, d\xi = \int e^{i(x-y)\xi/h} b_t((1-t)x+ty,\xi)\, d\xi,$$

where equality holds, e.g., in $\mathcal{D}'(\mathbf{R}_x^n \times \mathbf{R}_y^n)$. Setting

$$\begin{cases} \theta = x-y, \\ z = (1-t)x+ty = x-t\theta, \end{cases}$$

or, equivalently,

$$\begin{cases} x = z+t\theta, \\ y = z-(1-t)\theta, \end{cases}$$

we are led to

$$\int e^{i\theta\xi/h} b(z+t\theta, z-(1-t)\theta,\xi)\, d\xi = \int e^{i\theta\xi/h} b_t(z,\xi)\, d\xi,$$

where the right-hand side has to be interpreted as an oscillatory integral. In particular, by integrations by parts we see that it defines an element of $S_{2n}(1)$ with respect to the variables (z,θ). Since, moreover, the right-hand side is proportional to the inverse h-Fourier transform of $\xi \mapsto b_t(z,\xi)$, we obtain necessarily:

$$b_t(z,\zeta) = \frac{1}{(2\pi h)^n}\int e^{i(\xi-\zeta)\theta/h} b(z+t\theta, z-(1-t)\theta,\xi)\, d\xi\, d\theta, \tag{2.7.2}$$

where now the right-hand side has to be interpreted as a Fourier transform with respect to θ, and equality holds in $\mathcal{S}'(\mathbf{R}^{2n})$. Introducing again a cutoff function of the type $\chi(\theta, \xi - \zeta)$ (where χ is supported near zero), and making integrations by parts as in the proof of the theorem of composition, we see in the same way that $b_t \in S_{2n}(\langle\zeta\rangle^m)$, and the stationary phase theorem also gives

$$b_t(z,\zeta) \sim \sum_{\alpha\in\mathbf{N}^n} \frac{(-1)^{|\alpha|}h^{|\alpha|}}{i^{|\alpha|}\alpha!} \partial_\xi^\alpha \partial_\theta^\alpha (b(z+t\theta, z-(1-t)\theta, \xi)\Big|_{\substack{\theta=0\\ \xi=\zeta}}.$$

Finally, the unicity of b_t is a consequence of (2.7.2). ◇

Remark 2.7.2 If in (2.7.1) we make the change of variable $\theta \mapsto y = x - \theta$, we see that the formula can be rewritten as

$$b_t(x,\xi) = e^{-ix\xi/h}\mathrm{Op}_h\left(b(x+t(x-y), y+t(x-y), \xi)\right)\left(e^{i(\cdot)\xi/h}\right), \qquad (2.7.3)$$

where $e^{i(\cdot)\xi/h}$ stands for the function $y \mapsto e^{iy\xi/h}$. In particular, when $t = 0$ one can recover $\sigma_0(B)$ directly from B by

$$\boxed{\sigma_0(B)(x,\xi\,;\,h) = e^{-ix\xi/h}B\left(e^{i(\cdot)\xi/h}\right).} \qquad (2.7.4)$$

For other formulae concerning $\sigma_t(B)$, see Exercises 12 and 13 of this chapter.

Remark 2.7.3 Formula (2.7.1) can also be rewritten as

$$b_t(x,\xi;h) = e^{-ihD_\theta D_\xi} b(x+t\theta, x-(1-t)\theta, \xi)\,|_{\theta=0}\,. \qquad (2.7.5)$$

In particular, one can pass from any t-quantization to the t'-quantization by

$$\boxed{b_{t'}(x,\xi) = e^{-ih(t'-t)D_x D_\xi} b_t(x,\xi).} \qquad (2.7.6)$$

Examples

If $V = V(x) \in S_n(1)$, then for all $t \in [0,1]$ one has $\sigma_t(-h^2\Delta + V) = \xi^2 + V(x)$ which therefore does not depend on t. However, $\sigma_t(VhD_{x_1}) = V(x)\xi_1 + iht\partial_{x_1}V(x)$ while $\sigma_t(hD_{x_1}V) = V(x)\xi_1 + ih(t-1)\partial_{x_1}V(x)$.

An important feature of the previous reduction to symbols depending on $2n$ variables only, is that such symbols are uniquely determined by the pseudodifferential operator they quantize. The rule that makes it possible to determine the corresponding symbol of the composition of two pseudodifferential operators is the following:

Theorem 2.7.4 (Symbolic Calculus) *Let $a(x,\xi) \in S_{2n}(\langle\xi\rangle^m)$, and $b(x,\xi) \in S_{2n}(\langle\xi\rangle^{m'})$. Then for all $t \in [0,1]$ there exists $c_t \in S_{2n}(\langle\xi\rangle^{m+m'})$ (unique) such that*

$$\mathrm{Op}_h^t(a) \circ \mathrm{Op}_h^t(b) = \mathrm{Op}_h^t(c_t).$$

Moreover, c_t is given by

$$\boxed{c_t(x,\xi;h) = e^{ih[D_\eta D_v - D_u D_\xi]}\Big[a((1-t)x+tu,\eta)b(tx+(1-t)v,\xi)\Big]_{\substack{u=v=x\\ \eta=\xi}} =: a\#^t b,}\tag{2.7.7}$$

and it satisfies

$$c_t(x,\xi;h) \sim \sum_{k\geq 0} \frac{h^k}{i^k k!}(\partial_\eta\partial_v - \partial_\xi\partial_u)^k\Big[a((1-t)x+tu,\eta)b(tx+(1-t)v,\xi)\Big]_{\substack{u=v=x\\ \eta=\xi}}$$

in $S_{2n}(\langle\xi\rangle^{m+m'})$.

Proof By the theorem of composition, one has

$$\mathrm{Op}_h^t(a) \circ \mathrm{Op}_h^t(b) = \mathrm{Op}(c)$$

with

$$\begin{aligned} c(x,y,\xi) &= \frac{1}{(2\pi h)^n}\int e^{i(x-z)(\eta-\xi)/h} a((1-t)x+tz,\eta) \\ &\qquad\qquad \times b((1-t)z+ty,\xi)\,dz\,d\eta \end{aligned}\tag{2.7.8}$$

$$\sim \sum_{\alpha\in\mathbf{N}^n} \frac{h^{|\alpha|}}{i^{|\alpha|}\alpha!}\partial_z^\alpha\partial_\eta^\alpha\left(a((1-t)x+tz,\eta)b((1-t)z+ty,\xi)\right)_{\substack{z=x\\ \eta=\xi}}$$

in $S_{3n}(\langle\xi\rangle^{m+m'})$. Moreover, by the previous theorem,

$$\mathrm{Op}_h(c) = \mathrm{Op}_h^t(c_t)$$

with

$$\begin{aligned} c_t(x,\xi) &= \frac{1}{(2\pi h)^n}\int_{\mathbf{R}^{2n}} e^{i(\xi'-\xi)\theta/h} c(x+t\theta, x-(1-t)\theta,\xi')\,d\xi'\,d\theta \\ &\sim \sum \frac{(-1)^{|\alpha|}h^{|\alpha|}}{i^{|\alpha|}\alpha!}\partial_\xi^\alpha\partial_\theta^\alpha(c(x+t\theta,x-(1-t)\theta,\xi))_{\theta=0}, \end{aligned}\tag{2.7.9}$$

and therefore,

$$c_t(x,\xi) = \frac{1}{(2\pi h)^{2n}}\int e^{i(\xi'-\xi)\theta/h+(x+t\theta-z)(\eta-\xi')/h}a((1-t)(x+t\theta)+tz,\eta) \times b((1-t)z+t(x-(1-t)\theta),\xi')\,dz\,d\eta\,d\xi'\,d\theta.$$

Then we make the change of variables

$$(z,\theta)\mapsto(u,v):=(z+(1-t)\theta, z-t\theta),$$

which has determinant 1 and gives

$$c_t(x,\xi) = \frac{1}{(2\pi h)^{2n}}\int e^{i[x(\eta-\xi')+u(\xi'-\xi)-v(\eta-\xi)]/h}a((1-t)x+tu,\eta) \times b(tx+(1-t)v,\xi')\,dz\,d\eta\,d\xi'\,d\theta. \quad (2.7.10)$$

On the other hand, by the Fourier-inverse formula we have

$$e^{ih[D_\eta D_v - D_u D_\xi]}\Big[a((1-t)x+tu,\eta)b(tx+(1-t)v,\xi)\Big]_{\substack{u=v=x\\ \eta=\xi}}$$

$$= \frac{1}{(2\pi h)^{4n}}\int e^{i[(\xi-\eta)\eta^*+(x-v)v^*+(x-u)u^*+(\xi-\xi')\xi^*+\eta^*v^*-u^*\xi^*]/h} \times a((1-t)x+tu,\eta)b(tx+(1-t)v,\xi')\,du\,dv\,d\eta\,d\xi'\,du^*\,dv^*\,d\eta^*\,d\xi^*.$$

Since the integrations with respect to u^* and v^* give Dirac measures on $\{\xi^* = x-u\}$ and $\{\eta^* = v-x\}$, respectively, we get

$$e^{ih[D_\eta D_v - D_u D_\xi]}\Big[a((1-t)x+tu,\eta)b(tx+(1-t)v,\xi)\Big]_{\substack{u=v=x\\ \eta=\xi}}$$

$$= \frac{1}{(2\pi h)^{2n}}\int e^{i[(\xi-\eta)(v-x)+(\xi-\xi')(x-u)]/h}a((1-t)x+tu,\eta) \times b(tx+(1-t)v,\xi')\,du\,dv\,d\eta\,d\xi'.$$

Comparing this last formula with (2.7.10) we get immediately

$$c_t(x,\xi) = e^{ih[D_\eta D_v - D_u D_\xi]}\Big[a((1-t)x+tu,\eta)b(tx+(1-t)v,\xi)\Big]_{\substack{u=v=x\\ \eta=\xi}}.$$

Then the asymptotic expansion of c_t can be obtained either by again introducing cutoff functions and using the stationary phase theorem, or by composing directly the asymptotic expansions already obtained in (2.7.8) and (2.7.9) (see Exercise 11 at the end of this chapter). ◇

Particular Cases

For $t = 0$: The result could have been obtained more directly, and one obtains $\mathrm{Op}_h^0(a) \cdot \mathrm{Op}_h^0(b) = \mathrm{Op}_h^0(c^\ell)$, with

$$\boxed{c^\ell(x,\xi) = e^{ihD_\eta D_y} a(x,\eta) b(y,\xi)\Big|_{\substack{y=x\\ \eta=\xi}} =: a\#b \sim \sum_\alpha \frac{h^{|\alpha|}}{i^{|\alpha|}\alpha!} \partial_\xi^\alpha a(x,\xi)\, \partial_x^\alpha b(x,\xi).} \tag{2.7.11}$$

For $t = frac12$: We get $\mathrm{Op}_h^W(a) \cdot \mathrm{Op}_h^W(b) = \mathrm{Op}_h^W(c^W)$ with

$$\boxed{\begin{aligned} c^W(x,\xi) &= e^{ih[D_\eta D_x - D_y D_\xi]} a(y,\eta) b(x,\xi)\Big|_{\substack{y=x\\ \eta=\xi}} =: a\#^W b \\ &\sim \sum_{\alpha,\beta} \frac{h^{|\alpha+\beta|}(-1)^{|\alpha|}}{(2i)^{|\alpha+\beta|}\alpha!\beta!} (\partial_x^\alpha \partial_\xi^\beta a(x,\xi))(\partial_\xi^\alpha \partial_x^\beta b(x,\xi)). \end{aligned}} \tag{2.7.12}$$

Remark 2.7.5 If A and B are two pseudodifferential operators with symbols in $S_{2n}(\langle\xi\rangle^m)$ and $S_{2n}(\langle\xi\rangle^{m'})$, respectively, then for all $t \in [0,1]$, one has

$$\sigma_t(A \circ B) = \sigma_t(A)\sigma_t(B) + \mathcal{O}(h)$$

in $S_{2n}(\langle\xi\rangle^{m+m'})$.

Remark 2.7.6 : Case of a Commutator Writing $[A,B] = AB - BA$, the commutator of the two pseudodifferential operators A and B with symbols in $S_{2n}(\langle\xi\rangle^m)$ and $S_{2n}(\langle\xi\rangle^{m'})$, respectively, we get for all $t \in [0,1]$,

$$\sigma_t([A,B]) = \frac{h}{i}\{a,b\} + \mathcal{O}(h^2)$$

in $S_{2n}(\langle\xi\rangle^{m+m'})$, where $\{a,b\}$ denotes the *Poisson bracket* of a and b, defined by

$$\{a,b\} = \frac{\partial a}{\partial \xi}\frac{\partial b}{\partial x} - \frac{\partial a}{\partial x}\frac{\partial b}{\partial \xi},$$

or also, if we set

$$H_a = \frac{\partial a}{\partial \xi}\frac{\partial}{\partial x} - \frac{\partial a}{\partial x}\frac{\partial}{\partial \xi}$$

(which is called the *Hamilton field* of a),

$$\{a,b\} = H_a b = -H_b a.$$

Remark 2.7.7 : "Classical" symbols - A symbol $a \in S_{2n}(\langle\xi\rangle^m)$ is said to be *classical* if it admits an asymptotic expansion of the type

$$a(x,\xi;h) \sim \sum_{j\geq 0} h^j a_j(x,\xi)$$

with $a_j \in S_{2n}(\langle\xi\rangle^m)$ independent of h, and a_0 not identically zero. In this case, for any $\nu \in \mathbf{R}$, $h^\nu a_0(x,\xi)$ is called the *principal symbol* of the *classical* pseudodifferential operator $A = h^\nu \mathrm{Op}_h^t(a)$. By Theorem 2.7.1 we see that changing the quantization does not affect the "classical" character of a pseudodifferential operator, and moreover, its principal symbol does not depend on the quantization one uses. We set

$$h^\nu a_0 = \sigma_p(A).$$

In particular, by the two previous remarks, we have (in the case where the product $\sigma_p(A)\sigma_p(B)$ does not vanish identically)

$$\boxed{\sigma_p(AB) = \sigma_p(A)\sigma_p(B),}$$

and, if $\{\sigma_p(A), \sigma_p(B)\}$ does not vanish identically,

$$\boxed{\sigma_p([A,B]) = \frac{h}{i}\{\sigma_p(A), \sigma_p(B)\}.}$$

Also, using the formulae of Theorem 2.7.4, one can easily show the following, whose proof is left as an exercise:

Exercise (a)

If $a \in S_{2n}(\langle\xi\rangle^m)$ is classical, then there exists $b \in S_{2n}(\langle\xi\rangle^{-m})$ classical too, and $r = \mathcal{O}(h^\infty)$ in $S_{2n}(1)$, such that

$$\mathrm{Op}_h^t(a) \circ \mathrm{Op}_h^t(b) = 1 + \mathrm{Op}_h^t(r).$$

Remark 2.7.8 : ***Subprincipal Symbol*** If $a \sim \sum_{j\geq 0} h^j a_j$ is a classical symbol in $S_{2n}(\langle\xi\rangle^m)$, then a_1 is called the *subprincipal symbol* of $A = \mathrm{Op}_h^W(a)$.

Note that the Weyl quantization is required in this definition. In particular, using the formula of Theorem 2.7.1, one can show the following:

Exercise (b)

If $a \sim \sum_{j\geq 0} h^j a_j$ is a classical symbol in $S_{2n}(\langle\xi\rangle^m)$, then the subprincipal symbol b_1 of $\mathrm{Op}_h^0(a)$ is given by

$$b_1 = a_1 - \frac{1}{2i}\partial_x\partial_\xi a_0.$$

2.8 L^2-Continuity

Until now, we have made our pseudodifferential operators act on $\mathcal{S}(\mathbf{R}^n)$ and on $\mathcal{S}'(\mathbf{R}^n)$. However, for applications in quantum mechanics (where the physical states are described by functions in L^2), it is useful to know how these pseudodifferential operators transform $L^2(\mathbf{R}^n)$. The following result provides a rather complete answer to this problem:

Theorem 2.8.1 (Calderón–Vaillancourt) *Let $a \in S_{3n}(1)$. Then* $\mathrm{Op}(a)$ *is continuous on $L^2(\mathbf{R}^n)$, and* $\|\mathrm{Op}_h(a)\|_{\mathcal{L}(L^2)} \leq C_n \left(\sum_{|\alpha|\leq M_n} \|\partial^\alpha a\|_{L^\infty(\mathbf{R}^{3n})}\right)$, *where the positive constants C_n and M_n depend only on n.*

Proof By Theorem 2.7.1, there exists $b \in S_{2n}(1)$ such that $A = \mathrm{Op}_h^W(b)$. Moreover, using the operator $L := (1+\theta^2+(\xi-\xi')^2)^{-1}(1+h\theta D_{\xi'}+h(\xi'-\xi)D_\theta)$ to make integrations by parts in the integral expression of b, we see that for any $\alpha \in \mathbf{N}^{2n}$ the quantity $\|\partial^\alpha b\|_{L^\infty}$ can be estimated by a finite number of derivatives of a. As a consequence, it is enough to prove the theorem for $A = \mathrm{Op}_h^W(a)$ with $a \in S_{2n}(1)$. Moreover, by the change of variables $\xi \mapsto h\xi$ we see that

$$\mathrm{Op}_h^W(a(x,\xi)) = \mathrm{Op}_1^W(a(x,h\xi)),$$

and for all $\alpha, \beta \in \mathbf{N}^n$ we have

$$\partial_x^\alpha\partial_\xi^\beta(a(x,h\xi)) = h^{|\beta|}(\partial_x^\alpha\partial_\xi^\beta a)(x,h\xi) = \mathcal{O}(h^{|\beta|}\left\|\partial_x^\alpha\partial_\xi^\beta a\right\|_{L^\infty}).$$

Therefore, it is indeed enough to prove the result for $h = 1$, that is, for the operator

$$A = \mathrm{Op}_1^W(a).$$

We start by making a partition of $\mathbf{R}^{2n}$ thanks to the following lemma:

Lemma 2.8.2 *For all $d \in \mathbf{N}^*$ there exists $\chi_0 \in C_0^\infty(\mathbf{R}^d)$ such that, if we write $\chi_\mu(z) = \chi_0(z - \mu)$ (where $\mu \in \mathbf{Z}^d$), one has*

$$\sum_{\mu \in \mathbf{Z}^d} \chi_\mu = 1 \quad on \quad \mathbf{R}^d.$$

Proof Let $K = \left\{ z \in \mathbf{R}^d \,;\, |z_j| \leq \frac{1}{2} \text{ for } j = 1, \dots, d \right\}$. Then K is compact, and therefore there exists $\varphi \in C_0^\infty(\mathbf{R}^d)$ such that $\varphi \geq 0$ and $\varphi|_K = 1$. Set

$$\psi(z) = \sum_{\mu \in \mathbf{Z}^d} \varphi(z - \mu),$$

where actually, the sum runs over a finite set when z remains inside any fixed compact set. We have

$$\forall \nu \in \mathbf{Z}^d, \; \psi(z + \nu) = \psi(z)$$

and by construction,

$$\forall z \in \mathbf{R}^n, \; \psi(z) \geq 1.$$

Then $\chi_0 := \dfrac{\varphi}{\psi}$ solves the problem. ◇

Now we apply Lemma 2.8.2 with $d = 2n$, and for $\mu \in \mathbf{Z}^{2n}$ we set

$$a_\mu = a \chi_\mu.$$

Since $|\partial^\alpha \chi_\mu(z)| = |(\partial^\alpha \chi_0)(z - \mu)| \leq \sup |\partial^\alpha \chi_0|$ for all $\alpha \in \mathbf{N}^{2n}$, we obtain by the Leibniz formula

$$\partial^\alpha a_\mu = \mathcal{O}(\sup_{\beta \leq \alpha} \|\partial^\beta a\|_{L^\infty}) \text{ uniformly with respect to } \mu \in \mathbf{Z}^{2n}. \tag{2.8.1}$$

We set

$$A_\mu = \mathrm{Op}_1^W(a_\mu),$$

so that for any $u \in C_0^\infty(\mathbf{R}^n)$, one has

$$Au = \sum_\mu A_\mu u, \tag{2.8.2}$$

where the limit of the series can be taken, e.g., in $L^2(\mathbf{R}^n)$: Just make integrations by parts using the operator $L := (1 + |x - y|^2 + |\xi|^2)^{-1}(1 + h(x - y)D_\xi - h\xi D_y)$, and use the fact that on the support of $a_\mu\left(\frac{x+y}{2}, \xi\right) u(y)$ the quantity $|x - y| + |\xi|$ behaves like $\max\{|\mu|, |x| + |\xi|\}$ as $|(x, \xi)| \to +\infty$.

Now the key idea is to try to apply the following important lemma:

Lemma 2.8.3 (Cotlar–Stein Lemma) *Let $\mathcal{H}$ be a Hilbert-space, $(A_\mu)_{\mu\in\mathbf{Z}^p}$ a family of bounded operators on $\mathcal{H}$, and $\omega : \mathbf{Z}^d \to \mathbf{R}_+$ an application satisfying*

$$\forall \mu,\nu \in \mathbf{Z}^d, \qquad \|A_\mu A_\nu^*\| + \|A_\mu^* A_\nu\| \leq \omega(\mu-\nu)$$

and

$$C_0 := \sum_{\mu\in\mathbf{Z}^d} \sqrt{\omega(\mu)} < +\infty.$$

Then for all $M \geq 0$, one has,

$$\left\| \sum_{|\mu|\leq M} A_\mu \right\| \leq C_0.$$

We postpone for a moment the proof of this lemma, and continue with that of the Calderón–Vaillancourt theorem. For $\mu,\nu \in \mathbf{Z}^{2n}$, we have

$$A_\mu A_\nu^* u(x) = \int K_{\mu,\nu}(x,y)u(y)\,dy$$

with

$$K_{\mu,\nu}(x,y) = \frac{1}{(2\pi)^{2n}} \int e^{i(x\xi - y\eta - z\xi + z\eta)} a_\mu\left(\frac{x+z}{2},\xi\right) \bar{a}_\nu\left(\frac{y+z}{2},\eta\right) dz\,d\eta\,d\xi.$$

Since a_μ and a_ν are smooth and compactly supported, we see that $K_{\mu,\nu} \in C^\infty(\mathbf{R}^{2n})$, and we set

$$L = \frac{1}{1+|x-z|^2+|y-z|^2+|\xi-\eta|^2}(1+(x-z)\cdot D_\xi - (y-z)\cdot D_\eta - (\xi-\eta)\cdot D_z),$$

which satisfies

$$L\left(e^{i(x\xi-y\eta-z\xi+z\eta)}\right) = e^{i(x\xi-y\eta-z\xi+z\eta)}.$$

Using L to integrate by parts, we get for any $N \geq 0$,

$$\begin{aligned} K_{\mu,\nu}(x,y) &= \frac{1}{(2\pi)^{2n}} \int e^{i(x\xi-y\eta-z\xi+z\eta)} \\ &\qquad \times ({}^tL)^N \left(a_\mu\left(\frac{x+z}{2},\xi\right)\bar{a}_\nu\left(\frac{y+z}{2},\eta\right)\right) dy\,dz\,d\xi. \end{aligned}$$

Moreover, when $|\mu-\nu|$ is large enough, we have on $\mathrm{Supp}(a_\mu(t,\tau)\bar{a}_\nu(s,\sigma))$,

$$\frac{1}{C}|\mu-\nu| \leq |t-s| + |\tau-\sigma| \leq C|\mu-\nu|,$$

where $C > 0$ is a uniform constant. Setting $\mu = (\mu_1, \mu_2)$ and $\nu = (\nu_1, \nu_2)$ in $\mathbf{Z}^{2n}$, we deduce from this and (2.8.1) that

$$\int |K_{\mu,\nu}(x,y)|dy = \int_{\mathcal{D}_{\mu,\nu}} \mathcal{O}\left(\frac{\sup_{|\alpha|\leq N} \|\partial^\alpha a\|^2_{L^\infty}}{(1+|x-z|+|y-z|+|\xi-\eta|)^N}\right) dy\,dz\,d\eta\,d\xi,$$

where the $\mathcal{O}$ is uniform with respect to μ, ν, and a, and where

$$\mathcal{D}_{\mu,\nu} = \left\{\frac{1}{C}|\mu-\nu| \leq |x-y|+|\xi-\eta| \leq C|\mu-\nu|,\ |\xi-\mu_2| \leq C',\ |\eta-\nu_2| \leq C'\right\}$$

with $C' > 0$, which depends only on n.

Since $|x-z|+|y-z| \geq |x-y|$, this gives

$$\int |K_{\mu,\nu}(x,y)|dy = \int \frac{\mathcal{O}((1+|\mu-\nu|)^{2n+2-N} \sup_{|\alpha|\leq N} \|\partial^\alpha a\|^2_{L^\infty})}{(1+|x-z|)^{n+1}(1+|x-y|)^{n+1}} dy\,dz,$$

and therefore, for all $N \geq 0$,

$$\sup_{x\in\mathbf{R}^n} \int |K_{\mu,\nu}(x,y)|dy = \mathcal{O}\left((1+|\mu-\nu|)^{2n+2-N} \sup_{|\alpha|\leq N} \|\partial^\alpha a\|^2_{L^\infty}\right). \tag{2.8.3}$$

In the same way, we see that

$$\sup_{y\in\mathbf{R}^n} \int |K_{\mu,\nu}(x,y)|dx = \mathcal{O}\left((1+|\mu-\nu|)^{2n+2-N} \sup_{|\alpha|\leq N} \|\partial^\alpha a\|^2_{L^\infty}\right) \quad \forall N \geq 0. \tag{2.8.4}$$

The next step consists in applying the following elementary result:

Lemma 2.8.4 (Schur Lemma) *If $Au(x) = \int K(x,y)u(y)dy$ with $K \in C^0(\mathbf{R}^n \times \mathbf{R}^n)$, then*

$$\|A\|_{\mathcal{L}(L^2(\mathbf{R}^n))} \leq \left(\sup_x \int |K(x,y)|dy\right)^{1/2} \left(\sup_y \int |K(x,y)|dx\right)^{1/2},$$

where $\mathcal{L}(L^2(\mathbf{R}^n))$ denotes the space of bounded linear operators on $(L^2(\mathbf{R}^n))$.

Proof of the Schur Lemma Using the Cauchy–Schwarz inequality, we have

$$\begin{aligned} |Au(x)|^2 &\leq \left(\int |K(x,y)|^{1/2} \cdot |K(x,y)|^{1/2}|u(y)|dy\right)^2 \\ &\leq \int |K(x,y)|dy \cdot \int |K(x,y)| \cdot |u(y)|^2 dy, \end{aligned}$$

and therefore

$$\begin{aligned}\|Au\|^2 &\le \int \left(\int |K(x,y)|dy \cdot \int |K(x,y)| \cdot |u(y)|^2 dy\right) dx \\ &\le \sup_x \int |K(x,y)|dy \cdot \int\int |K(x,y)| \cdot |u(y)|^2 dy\, dx \\ &\le \sup_x \int |K(x,y)|dy \cdot \sup_y \int |K(x,y)|dx \cdot \|u\|^2.\end{aligned}$$

◇

Continuation of the proof of the theorem We deduce from the previous lemma and from (2.8.3)–(2.8.4) that for all $N \geq 0$,

$$\|A_\mu A_\nu^*\|_{\mathcal{L}(L^2)} = \mathcal{O}\left((1+|\mu-\nu|)^{2n+2-N}) \sup_{|\alpha|\le N} \|\partial^\alpha a\|_{L^\infty}^2\right)$$

uniformly with respect to μ and ν, and exactly in the same way one can prove

$$\|A_\mu^* A_\nu\|_{\mathcal{L}(L^2)} = \mathcal{O}\left((1+|\mu-\nu|)^{2n+2-N}) \sup_{|\alpha|\le N} \|\partial^\alpha a\|_{L^\infty}^2\right)$$

uniformly. Then we choose $N = 4n+3$, so that we can apply the Cotlar–Stein lemma with $d = 2n$ and

$$\omega(\mu) = C(1+|\mu|)^{-2n-1} \sup_{|\alpha|\le N} \|\partial^\alpha a\|_{L^\infty}^2,$$

where $C > 0$ depends only on n. We obtain that for any $M \geq 0$ and for any $u \in C_0^\infty$,

$$\Big\| \sum_{|\mu|\le M} A_\mu u \Big\|_{L^2} \le C_0 \|u\|_{L^2}$$

with

$$C_0 = \sqrt{C} \sum_{\mu\in \mathbf{Z}^{2n}} (1+|\mu|)^{-n-1/2} \sup_{|\alpha|\le N} \|\partial^\alpha a\|_{L^\infty}.$$

Taking the limit $M \to +\infty$, we get by (2.8.2),

$$\Big\|Au\Big\|_{L^2} \le C_0 \|u\|_{L^2},$$

and therefore, since u is arbitrary,

$$\|A\|_{\mathcal{L}(L^2)} \le C_0.$$

The proof of the theorem is complete, apart from the proof of the Cotlar–Stein lemma, which we give now:

Proof of the Cotlar–Stein Lemma Fix $M \geq 0$ and set $S = \sum_{|\mu| \leq M} A_\mu$. Then denoting by $\sigma(S^*S)$ the spectrum of the nonnegative self-adjoint operator S^*S, one has

$$\|S\|^2 = \sup_{\substack{u \in \mathcal{H} \\ \|u\|=1}} \|Su\|^2 = \sup_{\substack{u \in \mathcal{H} \\ \|u\|=1}} \langle S^*Su, u\rangle = \sup_{\lambda \in \sigma(S^*S)} |\lambda| = \|S^*S\|,$$

and therefore, for all $m \geq 1$,

$$\begin{aligned} \|S\|^{2m} &= \|S^*S\|^m = \left(\sup_{\lambda \in \sigma(S^*S)} |\lambda| \right)^m \\ &= \sup_{\lambda \in \sigma(S^*S)} |\lambda^m| = \sup_{\lambda' \in \sigma((S^*S)^m)} |\lambda'| \\ &= \|(S^*S)^m\|. \end{aligned}$$

On the other hand, we have

$$\begin{aligned} (S^*S)^m &= \left(\left(\sum_{|\mu| \leq M} A_\mu^* \right) \left(\sum_{|\nu| \leq M} A_\nu \right) \right)^m = \left(\sum_{|\mu|,|\nu| \leq M} A_\mu^* A_\nu \right)^m \\ &= \sum_{|\mu_\ell|,|\nu_\ell| \leq M} \underbrace{A_{\mu_1}^* A_{\nu_1} \ldots A_{\mu_m}^* A_{\nu_m}}_{B}. \end{aligned}$$

But, by assumption,

$$\|B\| \leq \|A_{\mu_1}^* A_{\nu_1}\| \ldots \|A_{\mu_m}^* A_{\nu_m}\| \leq \omega(\mu_1 - \nu_1) \ldots \omega(\mu_m - \nu_m)$$

and also, since $\|A_\mu\|^2 = \|A_\mu^* A_\mu\| \leq \omega(0)$,

$$\|B\| \leq \sqrt{\omega(0)}\, \omega(\nu_1 - \mu_2) \ldots \omega(\nu_{m-1} - \mu_m) \sqrt{\omega(0)}.$$

Taking the geometric mean of the two previous estimates, we get

$$\|B\| \leq \sqrt{\omega(0)} \sqrt{\omega(\mu_1 - \nu_1)} \sqrt{\omega(\nu_1 - \mu_2)} \ldots \sqrt{\omega(\mu_m - \nu_m)},$$

and therefore

$$
\begin{aligned}
&\|(S^*S)^m\| \\
&\leq \sum_{|\mu_1|\leq M}\sum_{|\nu_1|\leq M}\cdots\sum_{|\mu_m|\leq M}\sum_{|\nu_m|\leq M}\sqrt{\omega(0)}\sqrt{\omega(\mu_1-\nu_1)}\sqrt{\omega(\nu_1-\mu_2)}\ldots \\
&\qquad\qquad\ldots\sqrt{\omega(\mu_m-\nu_m)} \\
&\leq \sum_{|\mu_1|\leq M}\sum_{|\nu_1|\leq M}\cdots\sum_{|\mu_m|\leq M}\sqrt{\omega(0)}\sqrt{\omega(\mu_1-\nu_1)}\ldots\sqrt{\omega(\nu_{m-1}-\mu_m)}\cdot C_0 \\
&\qquad\ldots \\
&\leq \sum_{|\mu_1|\leq M}\sqrt{\omega(0)}C_0^{2m-1} \\
&\leq (2M+1)^n\sqrt{\omega(0)}C_0^{2m-1}.
\end{aligned}
$$

As a consequence,

$$\|S\| \leq ((2M+1)^n\sqrt{\omega(0)}C_0^{2m-1})^{1/2m}$$

for all $m \geq 1$. Taking the limit $m \to +\infty$, we get $\|S\| \leq C_0$. ◇

Consequences

(a) **Parametrix**: Using Proposition 2.6.10 and Theorem 2.8.1, we get that if $a \in S_{3n}(\langle\xi\rangle^m)$ is elliptic, there exists $b \in S_{3n}(\langle\xi\rangle^{-m})$ such that

$$\begin{cases} \mathrm{Op}_h(a)\circ\mathrm{Op}_h(b) = 1 + R_1, \\ \mathrm{Op}_h(b)\circ\mathrm{Op}_h(a) = 1 + R_2, \end{cases}$$

with

$$\|R_1\|_{\mathcal{L}(L^2)} + \|R_2\|_{\mathcal{L}(L^2)} = \mathcal{O}(h^\infty).$$

In particular, if $m = 0$ and h is small enough, $\mathrm{Op}_h(a)$ is invertible on $L^2(\mathbf{R}^n)$, with an inverse given by

$$
\begin{aligned}
\mathrm{Op}_h(a)^{-1} &= \mathrm{Op}_h(b)\left[\sum_{k=0}^{\infty}(-1)^k R_1^k\right] = \left[\sum_{k=0}^{\infty}(-1)^k R_2^k\right]\mathrm{Op}_h(b) \\
&= \mathrm{Op}_h(b) + \mathcal{O}(h^\infty).
\end{aligned}
$$

(b) **Gårding inequality**: Let $a \in S_{2n}(1)$ be real-valued and satisfy

$$a \geq \frac{1}{C}$$

for some positive constant C. Then we have the following result:

Proposition 2.8.5 *For every $C_1 > C$, one has*

$$\mathrm{Op}_h^W(a) \geq \frac{1}{C_1} \quad \text{on } L^2(\mathbf{R}^n)$$

if h is sufficiently small. Here the inequality is in the sense of operators and means that for all $u \in L^2(\mathbf{R}^n)$ one has $\left\langle \mathrm{Op}_h^W(a)u, u\right\rangle_{L^2} \geq \frac{1}{C_1}\|u\|_{L^2}^2$.

Proof Let $C_2 \in (C, C_1)$. Then $a - \frac{1}{C_2} \geq \frac{1}{C} - \frac{1}{C_2} > 0$, and therefore

$$\sqrt{a - \frac{1}{C_2}} \in S_{2n}(1).$$

Let $B = \mathrm{Op}_h^W\left(\sqrt{a - \frac{1}{C_2}}\right)$. Then B is bounded and self-adjoint on $L^2(\mathbf{R}^n)$, and by the symbolic calculus we have

$$\mathrm{Op}^W\left(a - \frac{1}{C_2}\right) = B^2 + hR$$

with $\|R\|_{\mathcal{L}(L^2)} = \mathcal{O}(1)$. In particular, since $B^2 \geq 0$, there exists a constant $C' > 0$ such that

$$\mathrm{Op}_h^W\left(a - \frac{1}{C_2}\right) \geq -C'h,$$

and therefore,

$$\mathrm{Op}_h^W(a) \geq \frac{1}{C_2} - C'h \geq \frac{1}{C_1}$$

for h small enough. ◇

2.9 A First Application: The Frequency Set

Although we shall develop more systematically this type of argument in the next sections (in the analytic framework), here we introduce a first notion of *microlocal behavior* of a function and we study its most basic properties. The word *microlocal* essentially means that the notion describes at the same time the local behaviors of both the function and its Fourier transform (see also Exercise 3 of Chapter 3 for another equivalent definition).

Definition 2.9.1 *Let $u \in L^2(\mathbf{R}^n)$ be an h-dependent function such that $\|u\|_{L^2} \leq 1$, and let $(x_0, \xi_0) \in \mathbf{R}^{2n}$. Then one says that u is* **microlocally infinitely small** *near (x_0, ξ_0) if there exists $\chi_0 \in S_{2n}(1)$ such that $\chi_0(x_0, \xi_0) = 1$ and*

$$\left\| \mathrm{Op}_h^W(\chi_0)u \right\|_{L^2} = \mathcal{O}(h^\infty)$$

uniformly as h tends to 0. The complement in $\mathbf{R}^{2n}$ of such points (x_0, ξ_0) is called the **frequency set** *of u and is denoted by* $\mathrm{FS}(u)$.

Remark 2.9.2 This notion was first introduced by V. Guillemin and S. Sternberg in [GuSt], and is the natural semiclassical analogue of the so-called *wave front set* (see [Ho2]) used to study the C^∞-singularities of a distribution.

Remark 2.9.3 It is easily seen that the condition $\chi_0(x_0, \xi_0) = 1$ can equivalently be replaced by $|\chi_0(x_0, \xi_0)| \geq \delta$ for some positive h-independent constant δ (in other words, the symbol χ_0 must be *elliptic* at (x_0, ξ_0)). Then it becomes obvious that $\mathrm{FS}(u)$ is a *closed* subset of $\mathbf{R}^{2n}$.

As a first consequence of this definition, we have the following proposition:

Proposition 2.9.4 *Let $u \in L^2(\mathbf{R}^n)$ such that $\|u\|_{L^2} \leq 1$, and let $(x_0, \xi_0) \notin \mathrm{FS}(u)$. Then there exists an open neighborhood V_0 of (x_0, ξ_0) in $\mathbf{R}^{2n}$ such that for* **any** *$\chi \in S_{2n}(1)$ supported in V_0, one has*

$$\left\| \mathrm{Op}_h^W(\chi)u \right\|_{L^2} = \mathcal{O}(h^\infty)$$

uniformly as h tends to 0.

Proof Let χ_0 be the symbol given in the definition. We first prove a lemma:

Lemma 2.9.5 *There exists $\chi_1 \in S_{2n}(1)$ such that $(x_0, \xi_0) \notin \mathrm{Supp}\chi_1$ and $\chi_0 + \chi_1$ is elliptic.*

Proof of the Lemma Since $\chi_0(x_0, \xi_0) = 1$ and $\nabla\chi_0$ is uniformly bounded, there exist two open neighborhoods V_0 and V_1 of (x_0, ξ_0) such that $\overline{V}_0 \subset V_1$ and

$$|\chi_0 - 1| \leq \frac{1}{3} \text{ on } V_0 \, ,$$
$$|\chi_0 - 1| \leq \frac{2}{3} \text{ on } V_1.$$

Then let $\varphi \in C^\infty(\mathbf{R}^{2n})$ such that $\varphi = 0$ on V_0, $\varphi = 1$ on $\mathbf{R}^{2n}\backslash V_1$, and $0 \leq \varphi \leq 1$ everywhere. Writing $M = \sup_{\mathbf{R}^{2n}} |\chi_0|$, we set $\chi_1 = 2M\varphi$. Then $\chi_1 \in S_{2n}(1)$, $V_0 \cap \text{Supp}\,\chi_1 = \emptyset$, and we have

- On $\mathbf{R}^{2n}\backslash V_1$, $|\chi_0 + \chi_1| \geq 2M - M = M \geq 1$;
- On V_1, $|\chi_0 + \chi_1| \geq \chi_1 + 1 - |\chi_0 - 1| \geq \frac{1}{3}$.

Therefore, χ_1 satisfies the properties required in the lemma. ◇

Now, denoting by $A = \text{Op}_h^W(a)$ a parametrix of $\text{Op}_h^W(\chi_0 + \chi_1)$ (with $a \in S_{2n}(1)$), we can write for any $\chi \in S_{2n}(1)$ supported in $V_0 = \mathbf{R}^{2n}\backslash\text{Supp}\,\chi_1$,

$$\text{Op}_h^W(\chi)u = \text{Op}_h^W(\chi)\left[\text{Op}_h^W(a)\text{Op}_h^W(\chi_0 + \chi_1) + R\right]u,$$

where $\|R\|_{\mathcal{L}(L^2)} = \mathcal{O}(h^\infty)$. Since $\left\|\text{Op}_h^W(\chi_0)u\right\| = \mathcal{O}(h^\infty)$ and $\|u\| \leq 1$, we deduce from this that

$$\text{Op}_h^W(\chi)u = \text{Op}_h^W(\chi)\text{Op}_h^W(a)\text{Op}_h^W(\chi_1)u + r = \text{Op}_h^W\left(\chi\#^W a\#^W\chi_1\right)u + r \tag{2.9.1}$$

with $\|r\|_{L^2} = \mathcal{O}(h^\infty)$. But since $\text{Supp}\,\chi \cap \text{Supp}\,\chi_1 = \emptyset$, we get from Theorem 2.7.4 that $\chi\#^W a\#^W\chi_1 = \mathcal{O}(h^\infty)$ in $S_{2n}(1)$. As a consequence, by the Calderón–Vaillancourt theorem the operator $\text{Op}_h^W\left(\chi\#^W a\#^W\chi_1\right)$ has norm $\mathcal{O}(h^\infty)$ on L^2, and therefore from (2.9.1) we get $\|\text{Op}_h^W(\chi)u\| = \mathcal{O}(h^\infty)$. ◇

The second result concerns the solutions of partial differential equations:

Proposition 2.9.6 *For some $m \in \mathbf{R}$, let $p \in S_{2n}(\langle\xi\rangle^m)$ be a classical symbol with principal part p_0. Moreover, let $u \in L^2(\mathbf{R}^n)$ with $\|u\|_{L^2} \leq 1$ be a solution of*

$$\text{Op}_h^W(p)u = 0.$$

Then

$$\mathrm{FS}(u) \subset p_0^{-1}(0) := \left\{(x,\xi) \in \mathbf{R}^{2n} \; ; \; p_0(x,\xi) = 0\right\}.$$

The set $p_0^{-1}(0)$ *is called the* **characteristic set** *of* $\mathrm{Op}_h^W(p)$ *(see also Definition 4.2.1 below).*

Proof Let (x_0,ξ_0) be such that $p_0(x_0,\xi_0) \neq 0$. Then there exists $\delta > 0$ such that for h small enough, $|p(x_0,\xi_0)| \geq \delta$. Moreover,

$$0 = \mathrm{Op}_h^W(\langle\xi\rangle^{-m})\mathrm{Op}_h^W(p)u = \mathrm{Op}_h^W\left(\langle\xi\rangle^{-m} \#^W p\right) u$$

and, writing $a = \langle\xi\rangle^{-m} \#^W p$, one has $a \in S_{2n}(1)$ and

$$a(x_0,\xi_0) = \langle\xi_0\rangle^{-m} p(x_0,\xi_0) + \mathcal{O}(h).$$

Therefore, $|a(x_0,\xi_0)| \geq \delta'$ for some other constant $\delta' > 0$ and for h small enough. As a consequence, $\chi_0(x,\xi) := a(x_0,\xi_0)^{-1}a(x,\xi)$ defines an element of $S_{2n}(1)$ that satisfies $\chi_0(x_0,\xi_0) = 1$ and $\mathrm{Op}_h^W(\chi_0)u = 0$. This proves that $(x_0,\xi_0) \notin \mathrm{FS}(u)$. ◇

Now let us show how information on $\mathrm{FS}(u)$ (that is, *microlocal* information) permits us in some cases to get *local* information on the behavior of u.

Proposition 2.9.7 *Assume that* $p \in S_{2n}(\langle\xi\rangle^m)$ *is such that there exists a constant* $C > 0$ *with*

$$|p(x,\xi)| \geq \frac{1}{C}\langle\xi\rangle^m \quad \text{in} \quad \mathbf{R}^n \times \{|\xi| \geq C\}.$$

(In this case, p is said to be "elliptic at infinity" in ξ*, or equivalently to be "elliptic in the classical sense".) Let* $u \in L^2(\mathbf{R}^n)$ *with* $\|u\|_{L^2} \leq 1$ *be a solution of*

$$\mathrm{Op}_h^W(p)u = 0$$

and assume that for some $x_0 \in \mathbf{R}^n$ *one has*

$$(\{x_0\} \times \mathbf{R}^n) \cap \mathrm{FS}(u) = \emptyset.$$

Then there exists a neighborhood W_0 *of* x_0 *in* $\mathbf{R}^n$ *such that*

$$\|u\|_{L^2(W_0)} = \mathcal{O}(h^\infty)$$

uniformly as h tends to 0.

Proof Let $\chi \in C_0^\infty(\mathbf{R}^n)$ such that $\chi(\xi) = 1$ for $|\xi| \le C$, and $0 \le \chi(\xi) \le 1$ everywhere. Then one has

$$\chi(\xi) + |p(x,\xi)|^2 \ge \frac{1}{C'} \langle\xi\rangle^{2m}$$

for some positive constant C' and for all $(x,\xi) \in \mathbf{R}^{2n}$. In particular, the symbol $\chi(\xi) + \left(\overline{p}\#^W p\right)(x,\xi)$ is elliptic in $S_{2n}(\langle\xi\rangle^{2m})$, and therefore there exists $a \in S_{2n}(\langle\xi\rangle^{-2m})$ such that

$$a\#^W \left(\chi(\xi) + \overline{p}\#^W p\right) = 1 + r$$

with $r = \mathcal{O}(h^\infty)$ in $S_{2n}(1)$. As a consequence, one has

$$\begin{aligned} u &= \mathrm{Op}_h^W(a)\mathrm{Op}_h^W(\chi(\xi))u + \mathrm{Op}_h^W(a)\mathrm{Op}_h^W(p)^*\mathrm{Op}_h^W(p)u - \mathrm{Op}_h^W(r)u \\ &= \mathrm{Op}_h^W(a)\mathrm{Op}_h^W(\chi(\xi))u + v \end{aligned}$$

with $\|v\|_{L^2} = \mathcal{O}(h^\infty)$. Now, since $\{x_0\} \times \mathrm{Supp}\chi$ is compact in $\mathbf{R}^{2n}$ and does not intersect $\mathrm{FS}(u)$, it is easy to deduce from Proposition 2.9.4 (e.g., by using a partition of unity) that there exists a neighborhood W_1 of x_0 and a neighborhood W_2 of $\mathrm{Supp}\chi$ such that for any $\varphi \in S_{2n}(1)$ supported in $W_1 \times W_2$ one has $\|\mathrm{Op}_h^W(\varphi)u\|_{L^2} = \mathcal{O}(h^\infty)$. Then let $\chi_1 = \chi_1(x) \in C_0^\infty(W_1)$ such that $\chi_1 = 1$ in a neighborhood W_0 of x_0. We have

$$\chi_1 u = \chi_1 \mathrm{Op}_h^W(a)\mathrm{Op}_h^W(\chi(\xi))u + \chi_1 v = \mathrm{Op}_h^W(\chi_1(x)\#^W a\#^W \chi(\xi))u + \chi_1 v,$$

and since $\chi_1(x)\#^W a\#^W\chi(\xi)$ is asymptotically equivalent to a symbol supported in $W_1 \times W_2$, we get that $\|\chi_1 u\| = \mathcal{O}(h^\infty)$. In particular, $\|u\|_{L^2(W_0)} = \mathcal{O}(h^\infty)$. ◇

Example: The Case of the Schrödinger Operator

The previous results can be applied to any normalized eigenfunction u of the semiclassical Schrödinger operator $P = -h^2\Delta + V$ with $V = V(x) \in S_n(1)$, assuming that the corresponding eigenvalue $E = E(h)$ admits a limit E_0 as h tends to 0. In this case we obtain in particular that for any compact $K \subset \{x \in \mathbf{R}^n\ ;\ V(x) > E_0\}$, one has $\|u\|_{L^2(K)} = \mathcal{O}(h^\infty)$ as h tends to 0. The region $\{x \in \mathbf{R}^n\ ;\ V(x) > E_0\}$ is called the *classically forbidden region.*

Remark 2.9.8 A result of propagation for $\mathrm{FS}(u)$ can also be proved (see Exercises 7 and 12 of Chapter 4) that in the particular case of the Schrödinger operator implies that $\mathrm{FS}(u)$ is a union of maximal classical trajectories.

2.10 Exercises and Problems

1. Prove that if $g \in C^\infty(\mathbf{R}^d\ ;\ \mathbf{R}_+^*)$ is such that $g(x) \geq e^{x^2}$ for all $x \in \mathbf{R}^d$, then g is not an order function. (Hint: Assume that $|\nabla g/g|$ $(= |\nabla \ln g|)$ is bounded, and get a contradiction.)

2. **Analytic Symbols** - A formal series $\sum_{j\in\mathbf{N}} h^j a_j(z)$ is called a *formal analytic symbol* on an open set $\Omega \subset \mathbf{C}^n$ if every a_j is holomorphic in Ω and there exists a constant $C_0 > 0$ such that for all $j \in \mathbf{N}$,

$$\sup_{z\in\Omega} |a_j(z)| \leq C_0^{j+1} j!. \tag{2.10.1}$$

(i) Prove that for any C_1 and $C_2 > 0$ large enough, there exists $\delta > 0$ such that

$$\sup_{z\in\Omega} \left| \sum_{0\leq j\leq \frac{1}{C_1 h}} h^j a_j(z) - \sum_{0\leq j\leq \frac{1}{C_2 h}} h^j a_j(z) \right| \leq e^{-\delta/h}$$

uniformly with respect to $h > 0$ small enough. (Hint: Use the Stirling formula: $j! \sim \sqrt{2\pi j}(j/e)^j$ for $j \to +\infty$.)

(ii) Assume that the formal series $\sum_{j\geq 0} h^j a_j(z)$ satisfies (formally) a differential equation in Ω of the type

$$\sum_{\substack{|\alpha|\leq m\\ 0\leq k\leq m}} h^k b_{k,\alpha}(z) D_z^\alpha \left(\sum_{j\geq 0} h^j a_j(z) \right) = 0$$

with the $b_{k,\alpha}$'s holomorphic in Ω. Then prove that for any open subset Ω_1 relatively compact in Ω and for any $C > 0$ large enough there exists $\delta > 0$ such that the function

$$a_C(z\ ;\ h) := \sum_{0\leq j\leq \frac{1}{Ch}} h^j a_j(z)$$

satisfies

$$\sup_{z\in\Omega_1} \left| \sum_{\substack{|\alpha|\leq m\\ 0\leq k\leq m}} h^k b_{k,\alpha}(z) D_z^\alpha a_C(z\ ;\ h) \right| \leq e^{-\delta/h}$$

uniformly with respect to $h > 0$ small enough. The function a_C is called an *analytic resummation* of the formal analytic symbol $\sum_{j\in\mathbf{N}} h^j a_j(z)$. Any symbol a that can be written as such a resummation up to an error $\mathcal{O}(e^{-\delta/h})$ for some $\delta > 0$ is called an *analytic symbol.* In this case it is also said that a is an *analytic representation* of the formal analytic symbol $\sum_{j\in\mathbf{N}} h^j a_j(z)$.

3. **Analytic Stationary Phase Theorem** - Let $a = a(x)$ be a bounded holomorphic function on some complex strip $\{x \in \mathbf{C} \; ; \; |\mathrm{Im} x| < \varepsilon\}$ with $\varepsilon > 0$, and consider the following oscillatory integral:

$$I(h) := \int_{\mathbf{R}} e^{ix^2/2h} a(x) dx.$$

(i) Using the operator $L := (1 + x^2)^{-1}(1 + hxD_x)$, write $I(h)$ as a convergent integral.

(ii) Setting

$$\widetilde{a}_h(x) := \left({}^tL\right)^2 a(x)$$

and observing that $I(h) = \sqrt{h} \int e^{-ih\xi^2/2} \mathcal{F}_1(\widetilde{a}_h)(\xi) d\xi$ (where $\mathcal{F}_1$ is the usual Fourier transform), show that for any $N > 0$ one has

$$I(h) = \sum_{k=0}^{N-1} \frac{\sqrt{2\pi h} h^k}{(2i)^k k!} (-1)^k \widetilde{a}_h^{(2k)}(0) + S_N$$

with

$$|S_N| \leq \frac{h^N}{N!} \int |\xi|^{2N} |\mathcal{F}_1(\widetilde{a}_h)(\xi)| d\xi.$$

(iii) By a change of contour of integration of the type: $x \mapsto x - i\delta\xi / \langle\xi\rangle$ ($\delta > 0$ small enough), show that there exists a constant $C > 0$ such that for all $h > 0$ small enough and for all $\xi \in \mathbf{R}$,

$$|\mathcal{F}_1(\widetilde{a}_h)(\xi)| \leq C e^{-\delta\xi^2/\langle\xi\rangle}.$$

(iv) Deduce from (ii) and (iii) that for some constant C_1, S_N satisfies

$$|S_N| \leq C_1^{N+1} N!$$

and then that $I(h)$ is an analytic resummation of a formal analytic symbol (Hint: Use Cauchy inequalities to show that the asymptotics found in (ii) define a formal analytic symbol, and take $N = 1/(C_2 h)$ with C_2 large enough.)

(v) Generalize the previous result to any dimension and for a more general quadratic phase of the type $\langle x, Qx\rangle$ with Q real symmetric.

4. **Pseudodifferential operators with analytic symbols** - Let $t \in [0,1]$ and $A = \mathrm{Op}_h^t(a)$ with $a = a(x,\xi\ ;\ h)$ a bounded analytic symbol on some complex strip $S_\varepsilon = \{|\mathrm{Im}x| + |\mathrm{Im}\xi| < \varepsilon\}$ ($\varepsilon > 0$ fixed).

   (i) Prove that for any $t' \in [0,1]$, $\sigma_{t'}(A)$ is asymptotically equivalent to a formal analytic symbol on $S_{\varepsilon'}$ for any $\varepsilon' < \varepsilon$. (Hint: Use formula (2.7.6) together with Cauchy estimates.)

   (ii) Using an argument similar to the one of the previous exercise, prove that $\sigma_{t'}(A)$ is indeed an analytic representation (as defined in Exercise 2) of its asymptotic series on $S_{\varepsilon'}$.

   (iii) Prove in the same way that if b is also a bounded analytic symbol on S_ε, then $a\#^t b$ is an analytic symbol on $S_{\varepsilon'}$ (see formula (2.7.7)).

5. **Gevrey symbols** - Do again the three previous exercises with the so-called *Gevrey symbols*, which are defined in the same way as the analytic symbols, except that the a_j's are assumed to be C^∞ only on some open set $\Omega \subset \mathbf{R}^n$, and the estimate (2.10.1) is now replaced by

$$\sup_{x\in\Omega} |\partial_x^\alpha a_j(x)| \leq C_0^{j+|\alpha|+1} (j!)^s (\alpha!)^s$$

   uniformly with respect to $\alpha \in \mathbf{N}^n$ and $j \in \mathbf{N}$. Here $s \in [1,+\infty)$ is some fixed number (the so-called *Gevrey index*). Verify that the case $s = 1$ corresponds to analytic symbols. (Hint: In the resummation, the quantity $1/Ch$ has to be replaced by $1/Ch^{1/s}$, leading to exponentially small errors of order $e^{-\delta/h^{1/s}}$.)

6. Generalize the WKB constructions made in Section 2.3 to the case where x_0 is in the *classically forbidden region*, that is, $V(x_0) > E$. (Hint: To get a remainder that is, at most $\mathcal{O}(h^\infty)$ there is only one possible choice for the phase $\varphi(x)$, since it must satisfy $\mathrm{Im}\varphi \geq 0$. Indeed, the remainder will then be $\mathcal{O}(h^\infty e^{-|\varphi(x)|/h})$.)

7. Generalize the WKB constructions made in Section 2.3 to the case of several dimensions, assuming that a smooth function $\varphi = \varphi(x)$ is given such that $|\nabla\varphi|^2 = E - V$. (Hint: Consider the flow on $\mathbf{R}^n$ generated by the vector field $\nabla\varphi$.)

8. **Airy Function** - For $\varepsilon, h > 0$, $x \in \mathbf{R}$, and $a = a(x,\xi) \in S_{2n}(1)$, one sets
$$I_a(x,h,\varepsilon) = \int_{\mathbf{R}} e^{-i(x\xi+\xi^3/3)/h} e^{-\varepsilon\langle\xi\rangle} a(x,\xi) d\xi.$$
(Here $\langle\xi\rangle := (1+\xi^2)^{1/2}$.)

   (i) Using the operator $L_1 = L_1(x,\xi,hD_\xi) = \dfrac{1}{2+\xi^2}(2-x-hD_\xi)$, show that $I_a(x,h,\varepsilon)$ tends, as $\varepsilon \to 0_+$, toward a function $u_a(x,h)$ that is, C^∞ with respect to $x \in \mathbf{R}$.

   (ii) Using the operator $L_2 = L_2(x,\xi,hD_\xi) = \dfrac{-h}{x+\xi^2} D_\xi$, show that for all $x > 0$, $u_a(x,h)$ is $\mathcal{O}(h^\infty)$.

   (iii) We assume from now on that $a = 1$ identically, and we denote by $u = u_1$ the corresponding function. Writing u as an oscillating integral, show that it is a solution of the following differential equation (the so-called *Airy equation*):
$$h^2 u''(x,h) = xu(x,h).$$

   (iv) In the integral that defines u, make the following change of contour of integration: $\xi \mapsto \xi - i$ (which will be justified by noticing that the coefficients of L_1 are holomorphic with respect to ξ in a neighborhood of $\{|\mathrm{Im}\xi| \le 1\}$). Then make the change of variable: $\xi = \sqrt{h}\eta$. Deduce that
$$u(1,h) \sim \sqrt{\pi h} e^{-2/3h} \qquad (h \to 0_+).$$

   (v) Now we are interested in the case $h = 1$. By a change of variable of the type $\xi = x^\alpha \eta$, show that the asymptotic behavior of $u(x,1)$ as $x \to +\infty$ can be deduced from that of $u(1,h)$ as $h \to 0_+$. Using (iv), find an equivalent of $u(x,1)$ as $x \to +\infty$.

   (vi) Find a simple equivalent of $u(x,1)$ as $x \to -\infty$. (Hint: Observe that for $x < 0$ the only critical points of $x\xi + \xi^3/3$ are real and nondegenerate.)

9. **The Heisenberg Uncertainty Principle** - Let $u \in H^1(\mathbf{R}^n)$ be such that $|x|u \in L^2(\mathbf{R}^n)$ and $\|u\|_{L^2} = 1$, and denote by $a = \langle xu, u\rangle$ and $b = \langle hD_x u, u\rangle$ its average position and its average momentum, respectively.

(i) Let $v(x) = e^{-ibx/h}u(x+a)$, and for $\lambda \in \mathbf{R}$ and $j \in \{1, \dots, n\}$, set

$$A_j(\lambda) = \int \left|\lambda x_j v(x) + iD_{x_j} v(x)\right|^2 dx.$$

Prove the identity

$$A_j(\lambda) = \|x_j v\|_{L^2}^2 \lambda^2 - \lambda + \|D_{x_j} v\|^2.$$

(ii) Using the fact that $A_j(\lambda) \geq 0$ for all $\lambda \in \mathbf{R}$, deduce from the previous identity that one has

$$\|(x_j - a_j)u\| \cdot \|(hD_{x_j} - b_j)u\| \geq \frac{h}{2}$$

for any $j = 1, \dots, n$. This estimate can be considered as a rigorous version of the Heisenberg uncertainty principle introduced in Chapter 1.

10. Prove that the asymptotic formula of the composition $A \circ B$ of two pseudodifferential operators becomes exact if A is a differential operator.

11. Prove the asymptotics of c_t in Theorem 2.7.4 by composing directly the asymptotic expansions obtained in (2.7.8) and (2.7.9). Hint: Notice that for any smooth function f one has

$$\begin{cases} \partial_\theta^\alpha(f(z,\theta)\,|_{z=x+t\theta}) = (t\partial_z + \partial_\theta)^\alpha f\;(x+t\theta,\theta), \\ \partial_\xi^\alpha(f(\eta,\xi)\,|_{\eta=\xi}) = (\partial_\eta + \partial_\xi)^\alpha f\;(\xi,\xi), \end{cases}$$

and use the change of variables:

$$\begin{cases} u = z + (1-t)\theta, \\ v = z - t\theta. \end{cases}$$

12. If $B = B(x, hD_x)$ is a pseudodifferential operator with symbol in $S(\langle\xi\rangle^m)$, prove that its t-symbol ($t \in [0,1]$) is given by the formula (the sense of which has to be justified)

$$b_t(x,\xi) = \frac{1}{(2\pi h)^n}\int e^{-i\xi\theta/h}\left[B(u,hD_u)\left(\delta_{\{u=x-(1-t)\theta\}}\right)\right]|_{u=x+t\theta}\,d\theta.$$

Hint: Use formula (2.7.1) or start directly from the right-hand side by writing $B(u, hD_u)$ explicitely.

13. With the same notation, using formulae (2.7.4) and (2.7.1) prove that

$$b_t(x,\xi) = \frac{1}{(2\pi h)^n}\int e^{i[(\xi'-\xi)\theta-(x+t\theta)\xi']/h}\left[B(u,hD_u)\left(e^{iu\xi'/h}\right)\right]|_{u=x+t\theta}\,d\xi' d\theta.$$

Compare with the previous formula.

14. Let $a \in S_{3n}(1)$ and set $P = hD_{x_n} + h\mathrm{Op}_h(a)$. Using the symbolic calculus, prove that there exists $b \in S_{2n}(1)$ elliptic such that

$$P\mathrm{Op}_h^W(b) = \mathrm{Op}_h^W(b)hD_{x_n} + R$$

with $\|R\|_{\mathcal{L}(L^2)} = \mathcal{O}(h^\infty)$.

15. If $a = a(x,\xi,h)$ satisfies

$$\partial_x^\alpha \partial_\xi^\beta a = \mathcal{O}(h^{-\mu|\alpha|-\nu|\beta|})$$

with $\mu+\nu \le 1$, then $\left\|\mathrm{Op}_h^W(a)\right\| = \mathcal{O}(1)$ uniformly with respect to $h > 0$. (Hint: Set $U_hf(x) = h^{n\mu/2}f(xh^\mu)$ and notice that $U_h\mathrm{Op}_h^W(a)U_h^{-1} = \mathrm{Op}_1^W(a(xh^\mu, \xi h^{1-\mu}))$.)

16. **Linear Symbols** - Let $L : \mathbf{R}^{2n} \to \mathbf{R}$ be a *linear* real-valued function on $\mathbf{R}^{2n}$.

    (i) Prove that $e^{iL} \in S_{2n}(1)$.

    (ii) Using formula (2.7.12), prove that for any $a \in S_{2n}(1)$ one has

    $$L\#^W a = La - \frac{ih}{2}\{L,a\}.$$

    (iii) Prove in the same way that for any other real-valued and linear function L' on $\mathbf{R}^{2n}$ such that $L\#^W L' = L'\#^W L$, one has

    $$e^{iL}\#^W e^{iL'} = e^{i(L+L')}.$$

    (iv) Setting $A_L = \mathrm{Op}_h^W(e^{iL})$, show that for all $u, v \in C_0^\infty(\mathbf{R}^n)$ one has

    $$\langle A_L u, v\rangle_{L^2} = \langle u, A_{-L}v\rangle_{L^2}$$

    and, using also (iii), deduce that A_L is a unitary operator on $L^2(\mathbf{R}^n)$.

(v) Let $u \in C_0^\infty(\mathbf{R}^n)$. For $t \in \mathbf{R}$ we set $\psi(t) = A_{tL}u$. Prove that $\psi \in C^1(\mathbf{R}_t\,;\, L^2(\mathbf{R}^n))$ and that ψ satisfies

$$\frac{\partial \psi}{\partial t} = i\widetilde{L}\psi, \quad \psi\,|_{t=0} = u,$$

with $\widetilde{L} := \mathrm{Op}_h^W(L) = L(x, hD_x)$. (This justifies the usual notation $A_{tL} = e^{itL(x,hD_x)}$, which can also be justified within the framework of spectral analysis.)

17. Let $a \in S_{2n}(1)$ such that its Fourier transform $\widehat{a} := \mathcal{F}_1 a$ is in $L^1(\mathbf{R}^{2n})$.

(i) With the notation introduced in the previous exercise, show that

$$\mathrm{Op}_h^W(a) = \frac{1}{(2\pi)^{2n}} \int \widehat{a}(x^*, \xi^*) e^{i(\langle x^*, x\rangle + \langle \xi^*, hD_x\rangle)} dx^* d\xi^*.$$

(ii) Using question (iv) of the previous exercise, deduce that the norm of $\mathrm{Op}_h^W(a)$ on $L^2(\mathbf{R}^n)$ satisfies

$$\|\mathrm{Op}_h^W(a)\|_{\mathcal{L}(L^2)} \leq \frac{1}{(2\pi)^{2n}} \|\widehat{a}\|_{L^1(\mathbf{R}^{2n})}.$$

18. Let $a \in S_{2n}(1) \cap L^1(\mathbf{R}^{2n})$. Show that for any $u, v \in C_0^\infty(\mathbf{R}^n)$ one has the formula

$$\left\langle \mathrm{Op}_h^W(a)u, v \right\rangle_{L^2} = \left(\frac{2}{\pi h}\right)^n \int e^{2iy\xi/h} a(x,\xi) u(x-y) v(x+y) dx\, dy\, d\xi,$$

and deduce the estimate:

$$\left\|\mathrm{Op}_h^W(a)\right\|_{\mathcal{L}(L^2)} \leq \left(\frac{2}{\pi h}\right)^n \|a\|_{L^1(\mathbf{R}^{2n})}.$$

19. **Hilbert–Schmidt Pseudodifferential Operators** - Let $a \in S_{2n}(1) \cap L^2(\mathbf{R}^{2n})$. Show that the distribution kernel $K_A(x,y)$ of $A = \mathrm{Op}_h^W(a)$ is in $L^2(\mathbf{R}^{2n})$ and satisfies

$$\|K_A\|_{L^2(\mathbf{R}^{2n})} = \frac{1}{(2\pi h)^{n/2}} \|a\|_{L^2(\mathbf{R}^{2n})}.$$

(Hint: Interpret $\|K_A\|^2$ as an oscillatory integral, make the change of variables $u = (x+y)/2$; $v = x - y$, and use the identity (2.4.5) to conclude.)

Note: Operators A that have a distribution kernel K_A in $L^2(\mathbf{R}^{2n})$ are called *Hilbert–Schmidt operators*, and the norm of K_A in L^2 is called the *Hilbert–Schmidt norm* of A and is denoted by $\|A\|_{\mathrm{HS}}$.

20. **Compact Pseudodifferential Operators** - Let $a \in S_{2n}(1)$ be such that for all $\alpha \in \mathbf{N}^{2n}$ one has

$$\partial^\alpha a(x,\xi) \to 0 \text{ as } |(x,\xi)| \to +\infty.$$

The purpose of this exercise is to show that for all $t \in [0,1]$ the operator $A_t := \mathrm{Op}_h^t(a)$ is compact on $L^2(\mathbf{R}^n)$.

(i) Using formula (2.7.1) prove that $A_t = \mathrm{Op}_h^0(b)$ with b satisfying

$$\partial^\alpha b(x,\xi) \to 0 \text{ as } |(x,\xi)| \to +\infty$$

for all $\alpha \in \mathbf{N}^{2n}$. (Hint: Make integrations by parts with the operator $L = (1+(\xi'-\xi)2+\theta^2)^{-1}(1+h(\xi'-\xi)D_\theta + h\theta D_{\xi'})$, and split the integral into several parts.)

(ii) Using the Calderón–Vaillancourt theorem, prove that for any $\varepsilon > 0$ there exists $\chi_\varepsilon \in C_0^\infty(\mathbf{R}^{2n})$ such that

$$\left\|\mathrm{Op}_h^0(b) - \mathrm{Op}_h^0(\chi_\varepsilon b)\right\| \leq \varepsilon.$$

(iii) Observing that $hD_x\mathrm{Op}_h^0(\chi_\varepsilon b) = \mathrm{Op}_h^0(\xi\chi_\varepsilon b + hD_x(\chi_\varepsilon b))$, prove that $\mathrm{Op}_h^0(\chi_\varepsilon b)$ is a bounded operator from $L^2(\mathbf{R}^n)$ to $H^1_{K_\varepsilon}(\mathbf{R}^n)$, where K_ε is the support of χ_ε and $H^1_{K_\varepsilon}(\mathbf{R}^n)$ denotes the space of functions in $H^1(\mathbf{R}^n)$ that are supported in K_ε.

(iv) Using that the natural injection $H^1_{K_\varepsilon}(\mathbf{R}^n) \to L^2(\mathbf{R}^n)$ is compact, deduce from (iii) that $\mathrm{Op}_h^0(\chi_\varepsilon b)$ is compact on $L^2(\mathbf{R}^n)$, and then from (ii) that the same is true for $\mathrm{Op}_h^0(b) = A_t$.

(v) Apply this result to the resolvent of the Schrödinger operator $R = (-h^2\Delta + V + i)^{-1}$, under the assumption that for some $k > 0$ the potential $V = V(x) \in S_n(\langle x\rangle^k$ is elliptic at infinity.

21. **Trace-Class Pseudodifferential Operators** - Let g be an order function on $\mathbf{R}^{2n}$ that in addition is in $L^1(\mathbf{R}^{2n})$. The purpose of this exercise

is to prove that for all $a \in S_{2n}(g)$ the operator $\mathrm{Op}_h^W(a)$ is trace-class on $L^2(\mathbf{R}^n)$ and its trace is given by

$$\mathrm{tr}\left(\mathrm{Op}_h^W(a)\right) = \frac{1}{(2\pi h)^n}\int a(x,\xi)dx\,d\xi.$$

Recall that an operator A is trace-class if it can be written as the product of two Hilbert–Schmidt operators, and that its trace is then given by $\mathrm{tr}(A) = \int K_A(x,x)dx$, where K_A is its distribution kernel.

(i) Assuming first that $a \in C_0^\infty(\mathbf{R}^{2n})$, prove directly the result in this case (one may, e.g., consider the operator $B = (1+h^2D_x^2+x^2)^{-(n+1)}$, which is Hilbert–Schmidt by Exercise 19, and write $A = B(B^{-1}A)$).

(ii) Using the result of Exercise 19, show that the exist $C_n > 0$ and $M_n \in \mathbf{N}$ such that for all $a \in C_0^\infty(\mathbf{R}^{2n})$ one has

$$\|\mathrm{Op}_h^W(a)\|_{tr} \leq C_n\left(1+\mathrm{diam}(\mathrm{Supp}a)\right)^{2n+2}\sum_{|\alpha|\leq M_n}\|\partial^\alpha a\|_{L^1(\mathbf{R}^{2n})},$$

where $\|\cdot\|_{tr}$ denotes the trace-class norm defined by

$$\|A\|_{tr} := \inf_{A_1A_2=A}\left(\|A_1\|_{HS}\cdot\|A_2\|_{HS}\right)$$

and diam(Supp$a$) stands for the diameter of the support of a. (Hint: Use the same decomposition as in the previous question, and observe that for any $(x_0,\xi_0) \in \mathbf{R}^{2n}$ one has

$$\|\mathrm{Op}_h^W(a)\|_{tr} = \|\mathrm{Op}_h^W(\tau_{(x_0,\xi_0)}a)\|_{tr}$$

with $\tau_{(x_0,\xi_0)}a(x,\xi) := a(x-x_0,\xi-\xi_0)$.)

(iii) Using the same partition of unity as in the proof of the Calderón–Vaillancourt theorem, deduce from (ii) that $\mathrm{Op}_h^W(a)$ is trace-class for all $a \in S_{2n}(g)$, and complete the proof.

22. **Sharp Gårding Inequality and Anti-Wick Quantization** - For $a \in S_{2n}(1)$, we set

$$\tilde{a}(x,\xi) = (\pi h)^{-n}\int a(y,\eta)e^{-[(x-y)^2+(\xi-\eta)^2]/h}dy\,d\eta.$$

(i) Show that $\tilde{a}$ is bounded, and then that $\tilde{a} \in S_{2n}(1)$ (one can observe that for any $\alpha \in \mathbf{N}^{2n}$ one has $\partial^\alpha \tilde{a} = \widetilde{\partial^\alpha a}$).

(ii) Prove that $\tilde{a} - a$ is $\mathcal{O}(h)$ in $S_{2n}(1)$. (Hint: Make a Taylor expansion of $a(y, \eta)$ of order 2 at the point (x, ξ).)

(iii) We define

$$\mathrm{Op}_h^{AW}(a) := \mathrm{Op}_h^W(\tilde{a})$$

(the so-called *anti-Wick quantization of* a). Deduce from (ii) and the Calderón–Vaillancourt theorem that

$$\left\| \mathrm{Op}_h^{AW}(a) - \mathrm{Op}_h^W(a) \right\|_{\mathcal{L}(L^2)} = \mathcal{O}(h).$$

(iv) Show that for any $u \in C_0^\infty$, one can write

$$\langle \mathrm{Op}_h^{AW}(a)u, u\rangle_{L^2} = \int a(y, \eta) e^{-y^2/h} |U(y, \eta)|^2 dy d\eta,$$

where $U(y, \eta)$ is a function that will have to be explicitly related to u.

(v) Deduce from (iii) and (iv) that if $a \geq 0$ everywhere, then there exists $C > 0$ such that for all $h \in (0, 1]$ one has

$$\mathrm{Op}_h^W(a) \geq -Ch;$$

that is, for all $u \in L^2(\mathbf{R}^n)$, $\langle \mathrm{Op}_h^W(a)u, u\rangle_{L^2} \geq -Ch\|u\|_{L^2}^2$ (the so-called *sharp Gårding inequality*).

23. **Almost Analytic Extensions** - For $f \in C^\infty(\mathbf{R}^n)$, we consider the formal series

$$S_f(x, y) = \sum_{\alpha \in \mathbf{N}^n} \frac{i^{|\alpha|}}{\alpha!} \partial^\alpha f(x) y^\alpha,$$

where $y \in \mathbf{R}^n$.

(i) By adapting the proof of Proposition 2.3.2, construct a function $\mathcal{A}_f \in C^\infty(\mathbf{R}^{2n})$ that is, asymptotically equivalent to S_f as $|y| \to 0$, in the sense that for all $N \in \mathbf{N}$ and $\beta, \gamma \in \mathbf{N}^n$ one has

$$\left| \partial_x^\beta \partial_y^\gamma \left(\mathcal{A}_f(x, y) - \sum_{|\alpha| \leq N} \frac{i^{|\alpha|}}{\alpha!} \partial^\alpha f(x) y^\alpha \right) \right| = \mathcal{O}(|y|^{(N - |\gamma|)_+})$$

locally uniformly with respect to $x \in \mathbf{R}^n$, and such that

$$\mathrm{Supp}\mathcal{A}_f \subset (\mathrm{Supp} f) \times K,$$

where K is a compact subset of $\mathbf{R}^n$.

(ii) Observing that $(\partial_x + i\partial_y)S_f(x,y) = 0$ in the sense of formal series, show that

$$\frac{1}{2}\left(\frac{\partial}{\partial x} + i\frac{\partial}{\partial y}\right)\mathcal{A}_f(x,y) = \mathcal{O}(|y|^\infty)$$

as $|y| \to 0_+$, locally uniformly with respect to x.

Since $(\partial_x + i\partial_y)/2$ is nothing but the $\overline{\partial}$ associated with the complex variable $z = x + iy$, this means that $\mathcal{A}_f(x,y)$ behaves more and more like a holomorphic function of z as $|y|$ becomes small. For this reason (and following the terminology of [MeSj]), $\mathcal{A}_f$ is called an *almost analytic extension* of f. Observe that when $f \in S_n(1)$, then $\mathcal{A}_f \in S_{2n}(1)$ and the estimates above become uniform with respect to $(x,y) \in \mathbf{R}^{2n}$ and h. If moreover $f \in C_0^\infty(\mathbf{R}^n)$, then $\mathcal{A}_f \in C_0^\infty(\mathbf{R}^{2n})$.

24. **Functional Calculus of Pseudodifferential Operators** - The purpose of this problem is to construct $f(A)$ as an h-pseudodifferential operator (modulo $\mathcal{O}(h^\infty)$) when A is a self-adjoint h-pseudodifferential operator and f is a smooth compactly supported function of $\mathbf{R}$. This construction is due to B. Helffer and J. Sjöstrand [HeSj5] and is based on the notion of almost analytic extension introduced in the previous exercise.

Let $a \in S_{2n}(1)$ be a real-valued symbol, and denote by $A = \mathrm{Op}_h^W(a)$ its semiclassical Weyl quantization. Let also $f \in C_0^\infty(\mathbf{R})$ and denote by $\mathcal{A}_f = \mathcal{A}_f(\mu,\nu) \in C_0^\infty(\mathbf{R}^2)$ an almost analytic extension of f (see the previous exercise).

(i) For $\lambda \in \mathbf{R}$ and $\varepsilon > 0$ we set

$$I_\varepsilon(\lambda) = \int_{|\nu|\geq\varepsilon} \frac{\overline{\partial}\mathcal{A}_f(\mu,\nu)}{(\mu + i\nu - \lambda)} d\mu d\nu,$$

where $\overline{\partial} := \frac{1}{2}\left(\frac{\partial}{\partial\mu} + i\frac{\partial}{\partial\nu}\right)$. Observing that $\overline{\partial}(\mu + i\nu - \lambda)^{-1} = 0$

and integrating by parts with respect to ν, prove that

$$I_\varepsilon(\lambda) = \frac{i}{2} \int \left(\frac{\mathcal{A}_f(\mu, -\varepsilon)}{\mu - i\varepsilon - \lambda} - \frac{\mathcal{A}_f(\mu, \varepsilon)}{\mu + i\varepsilon - \lambda} \right) d\mu.$$

(ii) Making a Taylor expansion to the first order of $\mathcal{A}_f(\mu, \pm\varepsilon)$ at $\varepsilon = 0$, show that

$$I_\varepsilon(\lambda) = -\pi f(\lambda) + \mathcal{O}(\varepsilon|\ln \varepsilon|)$$

uniformly with respect to $\lambda \in \mathbf{R}$.

(iii) Deduce that for any $\lambda \in \mathbf{R}$ one has the formula

$$f(\lambda) = -\frac{1}{\pi} \int_{\mathbf{R}^2} \frac{\overline{\partial} \mathcal{A}_f(\mu, \nu)}{(\mu + i\nu - \lambda)} d\mu d\nu.$$

(iv) In view of the previous formula, we set

$$\boxed{f(A) = -\frac{1}{\pi} \int_{\mathbf{R}^2} \overline{\partial} \mathcal{A}_f(\mu, \nu)(\mu + i\nu - A)^{-1} d\mu d\nu} \tag{2.10.2}$$

(this formula, due to Helffer and Sjöstrand [HeSj5], is related to older ideas by Dyn'kin [Dy] and has been shown to have applications in more general non-self-adjoint settings too, see, e.g., [Da, JeNa]). Using that $\|(z - A)^{-1}\| = \mathcal{O}(|\text{Im}\, z|^{-1})$ show that the integral in (2.10.2) converges in $\mathcal{L}(L^2)$, and that if f is real-valued, then the resulting bounded operator $f(A)$ is self-adjoint. (Indeed, because of the uniform convergence proved in (ii) it is easy to see that this operator coincides with the usual one defined by the spectral theorem for any self-adjoint operator A.)

(v) Using the symbolic calculus, construct for any $z \in \mathbf{C} \backslash \mathbf{R}$ a symbol $b_z \sim \sum_{j \geq 0} h^j b_{j,z}$ in $S_{2n}(1)$ such that

$$(z - a) \#^W b_z = 1 + \mathcal{O}(h^\infty) \text{ in } S_{2n}(1)$$

and show that for all $j \geq 0$ and $\alpha \in \mathbf{N}^{2n}$ one has

$$\partial^\alpha b_{j,z}(x, \xi) = \mathcal{O}(|\text{Im}\, z|^{-(2j+1+|\alpha|)})$$

as $\text{Im}\, z \to 0$, uniformly with respect to $(x, \xi) \in \mathbf{R}^{2n}$.

(vi) Deduce from (v) that for any $j \geq 0$ the function

$$c_j(x,\xi) := -\frac{1}{\pi}\int_{\mathbf{R}^2} \overline{\partial}\mathcal{A}_f(\mu,\nu) b_{j,\mu+i\nu}(x,\xi)\,d\mu\,d\nu$$

is in $S_{2n}(1)$, and show that $c_0(x,\xi) = f(a(x,\xi))$.

(vii) Denoting by $c \in S_{2n}(1)$ a resummation of the formal series $\sum_{j\geq 0} h^j c_j$, deduce from all the previous questions that

$$\|f(A) - \mathrm{Op}_h^W(c)\|_{\mathcal{L}(L^2)} = \mathcal{O}(h^\infty).$$

(viii) Generalize the previous construction to real-valued symbols $a \in S_{2n}(\langle\xi\rangle^{2k})$ such that $a + i$ is elliptic in $S_{2n}(\langle\xi\rangle^{2k})$. In particular, show that the result can be applied to the Schrödinger operator $A = -h^2\Delta + V(x)$ when $V \in S_n(1)$ is real-valued.

Note: Actually, better estimates can be obtained for the symbol of the resolvent $(z-A)^{-1}$: We refer the interested reader to the very nice book [DiSj], where further generalizations and applications are also given. The first proof that $f(A)$ is an h-pseudodifferential operator (up to $\mathcal{O}(h^\infty)$) is due to Helffer and Robert [HeRo].

Chapter 3

Microlocalization

3.1 The Global FBI Transform

For $u \in \mathcal{S}'(\mathbf{R}^n)$ we would like to treat simultaneously the local behavior of u and that of its h-Fourier transform $\mathcal{F}_h u$ (in this case, we say that we study the *microlocal* behavior of u).

For this purpose, we set

$$\begin{aligned} Tu(x,\xi;h) &= \underbrace{2^{-\frac{n}{2}}(\pi h)^{-\frac{3n}{4}}}_{=\alpha_{n,h}} \int e^{i(x-y)\xi/h-(x-y)^2/2h} u(y)\,dy \\ &\stackrel{\text{def}}{=} \alpha_{n,h} \left\langle u_y, e^{i(x-y)\xi/h-(x-y)^2/2h} \right\rangle_{\mathcal{S}',\mathcal{S}}, \end{aligned} \tag{3.1.1}$$

which belongs to $C^\infty(\mathbf{R}^{2n})$. Tu is called the *Fourier–Bros–Iagolnitzer* (for short, FBI) transform of u, and it has been used by many authors and for many purposes, in particular by J. Sjöstrand, who has also developed a systematic study for it in [Sj1, Sj2] (see also [Del]).

A possible explanation for this definition relies on the uncertainty principle (1.1.5): Here one tries to have $\Delta x \sim \Delta \xi \sim \sqrt{h}$, so we start by localizing u near x up to $\mathcal{O}\left(\sqrt{h}\right)$ by multiplying it with the Gaussian function $e^{-(x-y)^2/2h}$. Then one also tries to localize $\mathcal{F}_h u$ near ξ by just taking the Fourier transform with respect to y (the multiplication by $e^{ix\xi/h}$ is done only for the convenience of having a convolution operator). The fact that we have in this way localized $\mathcal{F}_h u$ near ξ up to $\mathcal{O}\left(\sqrt{h}\right)$ can be seen in the relation

$$Tu(x,\xi;h) = e^{ix\xi/h} T\mathcal{F}_h u(\xi,-x),$$

the proof of which is postponed (see Remark 3.4.4). As we shall see hereinafter in Proposition 3.1.1, the coefficient $\alpha_{n,h}$ is just a normalization factor.

Notice that all this can also be done by just doing a convolution with a Gaussian function, because one has

$$Tu(x,\xi;h) = \alpha_{n,h}e^{-\xi^2/2h}\int e^{-(x-i\xi-y)^2/2h}u(y)dy,$$

that is, setting $z = x - i\xi \in \mathbf{C}^n$, we have

$$Tu(x,\xi;h) = \alpha_{n,h}e^{-\xi^2/2h}\widetilde{T}u(z;h), \tag{3.1.2}$$

where $\widetilde{T}u(z;h) = \int e^{-(z-y)^2/2h}u(y)dy$ is called the *Bargman* transform of u. Also, writing $\phi_\xi(x) = (\pi h)^{-n/4}e^{ix\xi/h-x^2/2h}$ (the so-called *coherent state* centered at $(0,\xi)$, see also Exercise 1 of this chapter), one has (denoting by $*$ the usual convolution of functions)

$$Tu(x,\xi;h) = (2\pi h)^{-n/2}u * \phi_\xi(x),$$

and since $\mathcal{F}_h\phi_\xi(\eta) = (\pi h)^{-n/4}e^{-(\eta-\xi)^2/2h}$, this operation is also equivalent to the multiplication of $\mathcal{F}_hu$ by a Gaussian function.

The main elementary properties of this transform are as follows:

Proposition 3.1.1

(i) *For all $u \in \mathcal{S}'(\mathbf{R}^n)$, $e^{\xi^2/h}Tu(x,\xi;h)$ is a holomorphic function of $z = x - i\xi$ on $\mathbf{C}^n$.*

(ii) *If $u \in L^2(\mathbf{R}^n)$, then $Tu \in L^2(\mathbf{R}^{2n})$ and*

$$\|Tu\|_{L^2(\mathbf{R}^{2n})} = \|u\|_{L^2(\mathbf{R}^n)}$$

(that is, T maps $L^2(\mathbf{R}^n)$ isometrically into $L^2(\mathbf{R}^{2n})$.

(iii) *For all $u \in \mathcal{S}'(\mathbf{R}^n)$,*

$$hD_xTu = (\xi + ihD_\xi)Tu.$$

Proof

Assertion (i) is an immediate consequence of (3.1.2).

To prove (ii), it is enough to prove the equality for $u \in C_0^\infty(\mathbf{R}^n)$. Then for all $M, N > 0$ one has

$$\|Tu\|^2_{L^2\{|x|\leq M\,;\,|\xi|\leq N\}}$$
$$= \frac{1}{2^n(\pi h)^{3n/2}} \int_{\substack{|x|\leq M \\ |\xi|\leq N}} e^{i(y'-y)\xi/h-(x-y)^2/2h-(x-y')^2/2h} u(y)\overline{u(y')}dy\,dy'\,d\xi\,dx.$$

But, using, e.g., (2.4.5), we see that

$$\frac{1}{(2\pi h)^n}\int_{|\xi|\leq N} e^{i(y'-y)\xi/h}d\xi \underset{N\to+\infty}{\longrightarrow} \delta(y'-y) \text{ in } \mathcal{D}'(\mathbf{R}^{2n}_{y,y'}),$$

and thus

$$\|Tu\|^2_{L^2\left(\substack{|x|\leq M \\ |\xi|\leq N}\right)} \underset{N\to+\infty}{\longrightarrow} (\pi h)^{-n/2}\int_{|x|\leq M} e^{-(x-y)^2/h}|u(y)|^2dy\,dx.$$

Since we also have

$$\int_{\mathbf{R}^n} e^{-(x-y)^2/h}dx = \int_{\mathbf{R}^n} e^{-x^2/h}dx = h^{n/2}\int_{\mathbf{R}^n} e^{-x^2}dx = (\pi h)^{n/2},$$

we finally obtain, by the dominated convergence theorem,

$$\|Tu\|^2_{L^2(\mathbf{R}^{2n})} = \|u\|^2_{L^2(\mathbf{R}^n)},$$

from which the result follows by the density of $C_0^\infty(\mathbf{R}^n)$ in $L^2(\mathbf{R}^n)$.

Assertion (iii) is obtained immediately by differentiation under the summation sign. ◇

Remark 3.1.2 Still writing $z = x - i\xi$, one has

$$\partial_x - i\partial_\xi = 2\frac{\partial}{\partial \bar{z}},$$

so that actually (iii) can be seen as a consequence of (3.1.2) and (i).

Remark 3.1.3 As one can see in Exercise 2 at the end of this chapter, the image of $L^2(\mathbf{R}^n)$ by T is given by

$$T(L^2(\mathbf{R}^n)) = L^2(\mathbf{R}^{2n}) \cap e^{-\xi^2/2h}\mathcal{H}(\mathbf{C}^n_{x-i\xi}), \tag{3.1.3}$$

where $\mathcal{H}(\mathbf{C}^n_{x-i\xi})$ denotes the space of holomorphic functions with respect to $x - i\xi \in \mathbf{C}^n$.

An immediate consequence of (ii) is the following:

Corollary 3.1.4 *For all $u \in L^2(\mathbf{R}^n)$, one has $u = T^*Tu$.*

Remark 3.1.5 Of course, this does not mean that $T : L^2(\mathbf{R}^n) \to L^2(\mathbf{R}^{2n})$ is invertible. Indeed, Remark 3.1.3 proves that it is not, and actually, one can show (see Exercise 2 of this chapter) that TT^* is the orthogonal projector onto $L^2(\mathbf{R}^{2n}) \cap e^{-\xi^2/2h}\mathcal{H}(\mathbf{C}^n_{x-i\xi})$.

As another more general consequence of Proposition 3.1.1, we have the following:

Proposition 3.1.6 *T maps $\mathcal{S}'(\mathbf{R}^n)$ into $\mathcal{S}'(\mathbf{R}^{2n})$ continuously, and has its image included in $\mathcal{S}'(\mathbf{R}^{2n}) \cap C^\infty(\mathbf{R}^{2n})$. Moreover, for all $u \in \mathcal{S}'(\mathbf{R}^n)$ one has*

$$u = T^*Tu,$$

where for $v \in \mathcal{S}'(\mathbf{R}^{2n}) \cap C^\infty(\mathbf{R}^{2n})$ we have set

$$T^*v(y) = \alpha_{n,h} \int e^{-i(x-y)\xi/h-(x-y)^2/2h} v(x,\xi)\,dx\,d\xi,$$

which has to be interpreted as an oscillatory integral with respect to the ξ-variables.

Remark 3.1.7 In fact, one can prove (see Exercise 2 of this chapter) that

$$T(\mathcal{S}'(\mathbf{R}^n)) = \mathcal{S}'(\mathbf{R}^{2n}) \cap e^{-\xi^2/2h}\mathcal{H}(\mathbf{C}^n_{x-i\xi}).$$

Proof Since we already know that $T^*T = 1$ on $L^2(\mathbf{R}^n) \supset \mathcal{S}(\mathbf{R}^n)$, it is enough by duality to see that $T : \mathcal{S}(\mathbf{R}^n) \to \mathcal{S}(\mathbf{R}^{2n})$ and $T^* : \mathcal{S}(\mathbf{R}^{2n}) \to \mathcal{S}(\mathbf{R}^n)$ are continuous. Setting

$$L = \frac{1}{1+\xi^2}(1 - i\xi h D_y)$$

we have for all $N \in \mathbf{N}$,

$$Tu(x,\xi) = \alpha_{n,h} \int e^{i(x-y)\xi/h} ({}^tL)^N (e^{-(x-y)^2/2h} u(y))dy,$$

and therefore, since $({}^tL)^N$ is of order N in D_y and has coefficients that are $\mathcal{O}(\langle\xi\rangle^{-N})$, we get for all $\alpha, \beta \in \mathbf{N}^n$,

$$\partial_x^\alpha \partial_\xi^\beta Tu = \alpha_{n,h} \int e^{i(x-y)\xi/h-(x-y)^2/h} \mathcal{O}\left(\langle\xi\rangle^{|\alpha|-N} \langle x-y\rangle^{|\alpha|+|\beta|+N} \sum_{|\gamma|\leq N} |\partial^\gamma u(y)| \right) dy.$$

Now, using that

$$\left\{ \begin{array}{l} \langle x\rangle^k = \mathcal{O}(\langle y\rangle^k + \langle x-y\rangle^k), \\ \langle x-y\rangle^k e^{-(x-y)^2/2h} = \mathcal{O}(1), \end{array} \right.$$

for all $k \geq 0$, we get for any $k, k' \in \mathbf{N}$,

$$\begin{aligned} &\langle x\rangle^k \langle \xi\rangle^{k'} \partial_x^\alpha \partial_\xi^\beta Tu \\ &= \int \mathcal{O}\left(\underbrace{\langle \xi\rangle^{|\alpha|+k'-N}}_{=\mathcal{O}(1) \text{ if } N\gg 1} \underbrace{\langle x-y\rangle^{k+|\alpha|+|\beta|+N} e^{-(x-y)^2/2h}}_{\in L^1(\mathbf{R}^n_y)} \langle y\rangle^k \sum_{|\gamma|\leq N} |\partial^\gamma u(y)| \right) dy \\ &= \mathcal{O}\left(\sum_{|\gamma|\leq N} \sup_{y\in\mathbf{R}^n} \langle y\rangle^k |\partial^\gamma u(y)| \right) \end{aligned}$$

uniformly with respect to x and ξ, which proves that T maps $\mathcal{S}(\mathbf{R}^n)$ into $\mathcal{S}(\mathbf{R}^{2n})$ continuously.

The same type of arguments (but actually in a simpler way and without integration by parts) also show that $T^* : \mathcal{S}(\mathbf{R}^{2n}) \to \mathcal{S}(\mathbf{R}^n)$ is continuous, and this finishes the proof of Proposition 3.1.6. ◇

From the previous result, we see that if we know Tu on $\mathbf{R}^{2n}$, then we also know u. Moreover, since we have an explicit formula restoring u from Tu, we can also hope to derive properties of u just by knowing some properties of Tu. By definition, the local properties of Tu will be called *microlocal* properties of u.

For instance, the fact that $Tu = \mathcal{O}(h^\infty)$ near some point (x_0, ξ_0) will also be expressed by saying that u is *microlocally* $\mathcal{O}(h^\infty)$ near (x_0, ξ_0).

3.2 Microsupport

From now on, we are mainly interested in the exponential decay properties of Tu as h tends to 0. As we shall see, in some contexts (such as the semiclassical quantum mechanics) this will correspond to the exponential decay of u itself, but in another context (see [Sj1] and Remark 3.2.3 below), these properties of Tu are related to the (microlocal) analytic singularities of u.

Definition 3.2.1 *For $u \in \mathcal{S}'(\mathbf{R}^n)$ (h-dependent) and $(x_0, \xi_0) \in \mathbf{R}^{2n}$, we say that u is* **microlocally exponentially small** *near (x_0, ξ_0) if there exists some*

$\delta > 0$ such that

$$Tu(x, \xi; h) = \mathcal{O}(e^{-\delta/h})$$

uniformly for (x, ξ) in a neighborhood of (x_0, ξ_0), and $h > 0$ small enough. The complementary set of such points (x_0, ξ_0) is called the **microsupport** *of u, and is denoted $\mathrm{MS}(u) \subset \mathbf{R}^{2n}$.*

In other words, $\mathrm{MS}(u)$ is the subset of $\mathbf{R}^{2n}$ consisting of the points near which u is not microlocally exponentially small as $h \to 0$.

Remark 3.2.2 By definition, $\mathrm{MS}(u)$ is a *closed* subset of $\mathbf{R}^{2n}$.

Remark 3.2.3 In the case where u does not depend on h, $\mathrm{MS}(u)$ is closely related to the so-called *analytic wave front set* of u (see [Sj1]), which describes the microlocal analytic singularities of u. In fact, denoting by $\mathrm{WF}_a(u)$ this set, it is easy to prove that one then has (see Exercise 4 of Chapter 4)

$$\mathrm{MS}(u) = \mathrm{WF}_a(u) \cup [\mathrm{Supp}\, u \times \{0\}].$$

Remark 3.2.4 In the definition, we have used the L^∞-norm of Tu in a real neighborhood of (x_0, ξ_0). In fact, using assertion (i) of Proposition 3.1.1, we see that $Tu(x, \xi)$ can be extended to a holomorphic function on $\mathbf{C}^{2n}$, and for all $x, \xi, t, \tau \in \mathbf{R}^n$, we have

$$Tu(x + it, \xi + i\tau) = e^{t^2/2h + \tau^2/2h - \xi(t+i\tau)/h} Tu(x + \tau, \xi - t). \tag{3.2.1}$$

As a consequence, $|Tu|$ will be exponentially small in a real neighborhood of (x_0, ξ_0) if and only if it is so in a *complex* neighborhood of (x_0, ξ_0), and by the Cauchy formulae, this is again equivalent to the fact that $\|Tu\|_{L^p(\mathcal{V})}$ is exponentially small for some $p \geq 1$ and some complex neighborhood $\mathcal{V}$ of (x_0, ξ_0). But still using (3.2.1), this is also equivalent to the fact that $\|Tu\|_{L^p(\mathcal{W})}$ is exponentially small for some $p \geq 1$ and some *real* neighborhood $\mathcal{W}$ of (x_0, ξ_0). Therefore, in Definition 3.2.1, one can equivalently replace the local L^∞-norm of Tu by any local L^p-norm, $p \geq 1$.

As we shall see, in practice it is often more convenient to try to obtain L^2-type estimates on Tu. Then the previous discussion shows that they will automatically give uniform estimates on Tu (and as well on all the derivatives of Tu).

Moreover, in the applications it is also sometimes useful to try to localize more in the x-variables than in the ξ-ones, or vice versa. A possible way to do that is, to slightly modify the definition of T as follows: Fix $\mu > 0$, and for $u \in \mathcal{S}'(\mathbf{R}^n)$ set

$$\begin{aligned} T_\mu u(x,\xi;h) &= \mu^{\frac{n}{4}} 2^{-\frac{n}{2}} (\pi h)^{-\frac{3n}{4}} \int e^{i(x-y)\xi/h-\mu(x-y)^2/2h} u(y)dy \\ &= \mu^{-\frac{n}{2}} Tu\left(x, \frac{\xi}{\mu}; \frac{h}{\mu}\right). \end{aligned} \tag{3.2.2}$$

Then $T_1 = T$, $\|T_\mu u\|_{L^2} = \|u\|_{L^2}$, and when $\mu \to 0_+$, $\mu^{-n/4} T_\mu u(x,\xi;h)$ tends (e.g., in $\mathcal{S}'(\mathbf{R}^{2n})$) to $(\pi h)^{-n/4} e^{ix\xi/h} \mathcal{F}_h u(\xi)$, while when $\mu \to +\infty$, $\mu^{-3n/4}\, T_\mu u(x,\xi;h)$ tends to $\alpha_{n,h} u(x)$. As a consequence, for μ small $T_\mu u$ localizes more in ξ than in x, and the contrary holds for μ large.

Now the question is to know whether the previous definition of MS(u) is related to the special choice $\mu = 1$ that we have made. The answer is no, as stated in the following result:

Proposition 3.2.5 *Let $u \in \mathcal{S}'(\mathbf{R}^n)$. Then for all $\mu, \mu' > 0$ and $(x_0, \xi_0) \in \mathbf{R}^{2n}$ one has that $T_\mu u$ is exponentially small near (x_0,ξ_0) (as $h \to 0_+$) if and only if the same is true for $T_{\mu'} u$.*

Proof We are going to show that $T_\mu u$ is exponentially small if and only if the same is true for $T_1 u$. Assume first that for some $\delta > 0$, $T_1 u = \mathcal{O}\left(e^{-\delta/h}\right)$ in a neighborhood $\mathcal{V}_0$ of (x_0,ξ_0). We know that $u = T_1^* T_1 u$, and thus

$$\begin{aligned} T_\mu u(x,\xi) &= (T_\mu T_1^*) T_1 u(x,\xi) \\ &= \mu^{\frac{n}{4}} \alpha_{n,h}^2 \int e^{(2i(x-y)\xi - \mu(x-y)^2 - 2i(z-y)\zeta - (z-y)^2)/2h} T_1 u(z,\zeta) dz\, d\zeta\, dy \\ &= \mu^{\frac{n}{4}} \alpha_{n,h}^2 \int e^{(2iy(\zeta-\xi) + 2i(x\xi - z\zeta) - \mu(x-y)^2 - (z-y)^2)/2h} T_1 u(z,\zeta) dz\, d\xi\, dy. \end{aligned}$$

Now,

$$\begin{aligned} \mu(x-y)^2 + (z-y)^2 &= \mu x^2 + z^2 + (1+\mu)y^2 - 2y(\mu x + z) \\ &= (1+\mu)\left(y - \frac{\mu x + z}{1+\mu}\right)^2 + \frac{\mu}{1+\mu}(x-z)^2 \end{aligned}$$

and

$$\begin{aligned} \int e^{iy(\zeta-\xi)/h - (1+\mu)(y + \frac{\mu x + z}{1+\mu})^2/2h} dy &= e^{-i\frac{\mu x + z}{1+\mu}(\zeta-\xi)/h} \int e^{iy(\zeta-\xi)/h - (1+\mu)y^2/2h} dy \\ &= \left(\frac{2\pi h}{1+\mu}\right)^{\frac{n}{2}} e^{-i\frac{\mu x+z}{1+\mu}(\zeta-\xi)/h} e^{-(\xi-\zeta)^2/2(1+\mu)h}. \end{aligned}$$

Therefore,

$$T_\mu u(x,\xi) = \left(\frac{2\pi h\sqrt{\mu}}{1+\mu}\right)^{\frac{n}{2}} \alpha_{n,h}^2 \int e^{i(x\xi - z\zeta)/h - i\frac{\mu x + z}{1+\mu}(\zeta-\xi)/h} \times e^{-\frac{\mu}{1+\mu}(x-z)^2/2h - \frac{1}{(1+\mu)}(\xi-\zeta)^2/2h} T_1 u(z,\zeta)\, dz\, d\zeta,$$

and taking $\chi \in C_0^\infty(\mathcal{V}_0)$ such that $\chi = 1$ near (x_0, ξ_0), we can write

$$T_\mu u(x,\xi) = A(x,\xi) + B(x,\xi)$$

with

$$\begin{aligned} A(x,\xi) &= \left(\frac{2\pi h\sqrt{\mu}}{1+\mu}\right)^{\frac{n}{2}} \alpha_{n,h}^2 \int \chi(z,\zeta) e^{i(x\xi - z\zeta)/h - i\frac{\mu x + z}{1+\mu}(\zeta-\xi)/h} \\ &\quad \times e^{-\frac{\mu}{1+\mu}(x-z)^2/2h - \frac{1}{(1+\mu)}(\xi-\zeta)^2/2h} T_1 u(z,\zeta)\, dz\, d\zeta \\ &= \mathcal{O}(e^{-\delta/h}), \end{aligned}$$

where the last equality comes from the assumption we have made on $T_1 u$, and the fact that

$$h^{\frac{n}{2}} \alpha_{n,h}^2 \int e^{-\frac{\mu}{1+\mu}(x-z)^2/2h - \frac{1}{(1+\mu)}(\xi-\zeta)^2/2h} dz\, d\zeta = \mathcal{O}(1)$$

uniformly with respect to h. Moreover, if $(z,\zeta) \in \mathrm{Supp}(1-\chi)$, then $(x_0 - z)^2 + (\xi_0 - \zeta)^2 \geq \frac{1}{C}$ for some positive constant C, and therefore $(x-z)^2 + (\xi-\zeta)^2 \geq \frac{1}{2C}$ if (x,ξ) is sufficiently close to (x_0, ξ_0). As a consequence, we see that

$$B(x,\xi) = \mathcal{O}(e^{-\delta'/h})$$

with, e.g., $\delta' = \dfrac{\min(\mu, 1)}{5(1+\mu)C} > 0$. Therefore, we get that $T_\mu u(x,\xi)$ is exponentially small near (x_0, ξ_0).

Conversely, the result follows in the same way by writing $T_1 u = (T_1 T_\mu^*) T_\mu u$.

◇

Remark 3.2.6 Actually, the previous proof also shows that if $T_1 u$ satisfies an estimate of the type

$$\|T_1 u(x,\xi;h)\|_{L^2(\mathcal{V})} \leq r(h),$$

where $\mathcal{V}$ is a neighborhood of (x_0, ξ_0), then there exists $\delta > 0$ such that

$$\|T_\mu u(x,\xi;h)\|_{L^2(\mathcal{V}')} \leq \frac{2^{\frac{n}{2}}(1+\mu)^{\frac{n}{2}}}{\mu^{\frac{n}{4}}} r(h) + e^{-\delta/h}$$

uniformly for h small enough, where $\mathcal{V}'$ is a (possibly smaller) neighborhood of (x_0, ξ_0). As a consequence, we have a similar invariance for the analogues of MS(u) that are obtained by replacing the decay $\mathcal{O}(e^{-\delta/h})$ for the local L^2-norm of $T_1 u$ by $\mathcal{O}(h^\infty)$, $\mathcal{O}(h^s)$ or $\mathcal{O}(e^{-\delta/h^{1/\alpha}})$ ($s \in \mathbf{R}$ and $\alpha \geq 1$ fixed). These sets are respectively denoted by FS(u) (the *frequency set* of u: see also Section 2.9 and Exercise 3 at the end of this chapter), FS$^{(s)}(u)$, and MS$_\alpha(u)$. When u does not depend on h, they are associated with the microlocal C^∞ (respectively H^s and G_α) singularities of u, where H^s is the usual Sobolev space of order s, and G_α is the Gevrey space of order α.

Of course, other notions of microsupports or frequency sets can be considered, by modifying the choice of the local decay of $T_1 u$ as $h \to 0_+$. To obtain an invariant definition, however, it is necessary for this decay to be at most exponential.

Remark 3.2.7 The notion of MS(u) is *local*, in the sense that if u and v are two tempered distributions that coincide on some open set $\Omega \subset \mathbf{R}^n$, then MS$(u) \cap (\Omega \times \mathbf{R}^n) =$ MS$(v) \cap (\Omega \times \mathbf{R}^n)$. This is an easy consequence of the presence of the Gaussian localization factor $e^{-(x-y)^2/2h}$ in the definition of T.

3.3 Action of the FBI Transform on ΨDOs

As we shall see in this section, a very pleasant property of the global FBI transform is that it transforms in a very explicit way the pseudodifferential operators (for short, ΨDOs) on $\mathbf{R}^n$ into pseudodifferential operators on $\mathbf{R}^{2n}$. This will be very useful for getting information on the solutions of partial differential equations (in particular, on their microsupport), since the combination of the following result with that of the next section will permit us to relate their properties to the geometric ones of the symbol of the equation.

For any symbol $p \in S_{2n}(1)$ as defined in Chapter 2, we have the following result:

Proposition 3.3.1 *For any* $t \in [0,1]$, *one has*

$$T \circ \mathrm{Op}_h^t(p) = \mathrm{Op}_h^t(\tilde{p}) \circ T,$$

where $\widetilde{p} \in S_{4n}(1)$ *is defined by*

$$\widetilde{p}(x,\xi,x^*,\xi^*) = p(x-\xi^*,x^*).$$

Here x^* *and* ξ^* *denote the dual variables of* x *and* ξ, *respectively, so that* $\mathrm{Op}(x^*) = hD_x$ *and* $\mathrm{Op}(\xi^*) = hD_\xi$.

Remark 3.3.2 As one can notice, this formula is exact (that is, without any smaller remainder term), and the symbol $\widetilde{p}$ that is, obtained does not depend on the choice of the quantization (i.e., on t).

Proof For $u \in C_0^\infty(\mathbf{R}^n)$, we have

$$\begin{aligned}
&\mathrm{Op}_t(p(x-\xi^*,x^*))Tu(x,\xi) &(3.3.1)\\
&= \frac{\alpha_{n,h}}{(2\pi h)^{2n}}\int_{\mathbf{R}^{5n}} e^{i\Phi/h}p((1-t)x+tx'-\xi^*,x^*)u(y)dy\,dx'\,d\xi'\,dx^*\,d\xi^*
\end{aligned}$$

with

$$\Phi = (x-x')x^* + (\xi-\xi')\xi^* + (x'-y)\xi' + i(x'-y)^2/2.$$

Then integrating first with respect to ξ' and using the fact that

$$\int e^{i(x'-y-\xi^*)\xi'/h}d\xi' = (2\pi h)^n\delta_{\xi^*=(x'-y)}$$

we get from (3.3.1),

$$\begin{aligned}
&\mathrm{Op}_t(p(x-\xi^*,x^*))Tu(x,\xi) &(3.3.2)\\
&= \frac{\alpha_{n,h}}{(2\pi h)^{n}}\int e^{i\Phi_1/h}p((1-t)(x-x')+y,x^*)u(y)dy\,dx'\,dx^*
\end{aligned}$$

with

$$\Phi_1 = (x'-y)\xi + (x-x')x^* + i(x'-y)^2/2.$$

Finally, making the change of variables $x' \mapsto z = x - x' + y$ in (3.3.2), we obtain

$$\mathrm{Op}_t(p(x-\xi^*,x^*))Tu(x,\xi) = \frac{\alpha_{n,h}}{(2\pi h)^{n}}\int e^{i\Phi_2/h}p((1-t)z+ty,x^*)u(y)dy\,dz\,dx^*$$

with

$$\Phi_2 = (x-z)\xi + i(x-z)^2/2 + (z-y)x^*$$

and therefore

$$\mathrm{Op}_t(p(x-\xi^*,x^*))Tu(x,\xi) = T(\mathrm{Op}_h^t(p)u)(x,\xi).$$

◇

3.4 Action of the FBI Transform on FIOs

In Remark 2.5.2 we already have had a taste of Fourier integral operators (for short, FIOs). In this paragraph we consider a special kind of FIOs, namely those associated with *linear canonical transformations* in a sense that will become clear only in Chapter 5 (see in particular Exercise 1 of Chapter 5): Here we must stress the fact that this section does not contain results used in the sequel, and it can therefore be skipped at a first reading. However, it is interesting to note that the considerations of this section always lead to completely explicit and exact formulae (such FIOs are also related to the so-called *exact Egorov theorem*, an example of which is given in Exercise 10 of Chapter 4). Moreover, these results can be useful in some problems requiring local canonical changes of variables. For similar considerations one may also consult [Fo].

In order to describe the action of the FBI transform on FIOs, we need to generalize again the class of FBI transforms we work with. If $A = A_1 + iA_2$ is a symmetric $n \times n$ matrix such that $A_1 = (A + \overline{A})/2$ is positive definite, we set, for $x \in \mathbf{R}^n$,

$$q_A(x) = \langle Ax, x \rangle ,$$

and we set, for $u \in \mathcal{S}'(\mathbf{R}^n)$,

$$T_A u(x, \xi; h) = (\det A_1)^{\frac{1}{4}} \alpha_{n,h} \int e^{i(x-y)\xi/h - q_A(x-y)/2h} u(y) dy. \tag{3.4.1}$$

Then we see that

$$T_A u(x, \xi; h) = e^{i\langle A_2 x, x\rangle/2h} T_{A_1}(e^{-i\langle A_2 y, y\rangle/2h} u)(x, \xi - A_2 x)$$

and

$$T_{A_1} u(x, \xi; h) = (\det A_1)^{-\frac{1}{4}} T(u \circ A_1^{-\frac{1}{2}})(A_1^{\frac{1}{2}} x, A_1^{-\frac{1}{2}} \xi; h)$$

(where $T = T_I$ is the usual FBI transform defined previously). As a consequence T_A is an isometry from $L^2(\mathbf{R}^n)$ to $L^2(\mathbf{R}^{2n})$, too.

At first, we consider the three following types of (unitary) Fourier integral operators (next, we shall anyway limit our study to FIOs with real quadratic phase):

Type 1:

$$J_B \; : \; u \mapsto |\det B|^{\frac{1}{2}} (u \circ B),$$

where B is an invertible $n \times n$ matrix.

Type 2:

$$K_C \,:\, u \mapsto K_C u(x) = e^{-i\langle Cx,x\rangle/2h} u(x),$$

where C is a real symmetric $n \times n$ matrix.

Type 3:

$$L_j \,:\, u \mapsto L_j u(x_1, \ldots, x_{j-1}, \xi_j, x_{j+1}, \ldots, x_n) = \frac{1}{\sqrt{2\pi h}} \int e^{-ix_j\xi_j/h} u(x) dx_j,$$

where $j \in \{1, \ldots, n\}$.

With these operators we associate respectively the three following transformations on $\mathbf{R}^{2n}$:

Type 1:

$$j_B \,:\, (x, \xi) \mapsto (Bx, {}^tB^{-1}\xi);$$

Type 2:

$$k_C \,:\, (x, \xi) \mapsto (x, \xi + Cx);$$

Type 3:

$$\ell_j \,:\, (x, \xi) \mapsto ((x_1, \ldots, x_{j-1}, -\xi_j, x_{j+1}, \ldots, x_n), (\xi_1, \ldots, \xi_{j-1}, x_j, \xi_{j+1}, \ldots, \xi_n)).$$

As is easy to verify, these transformations have the property of leaving unchanged the so-called *canonical 2-form* (or *canonical symplectic form*) σ on $\mathbf{R}^{2n}$, defined by

$$\sigma((x, \xi), (y, \eta)) = \xi y - x\eta,$$

in the sense that if κ denotes any one of them, it satisfies

$$\sigma(\kappa(X), \kappa(Y)) = \sigma(X, Y) \tag{3.4.2}$$

for all $X, Y \in \mathbf{R}^{2n}$. For this reason, such transformations are called *canonical* or *symplectic* transformations (see also Chapter 5 for more general considerations about this type of transformations). Moreover, it can be shown that the group of all the linear symplectic transformations on $\mathbf{R}^{2n}$ (i.e. satisfying (3.4.2)) is generated by those belonging to the three previous types; that is, any linear symplectic transformation on $\mathbf{R}^{2n}$ can be written as the composition of a finite number of j_B's, k_C's, and ℓ_j's (see, e.g., [Fo], Proposition (4.10)).

If κ is any linear canonical transformation on $\mathbf{R}^{2n}$, we define for $(x,\xi)\in\mathbf{R}^{2n}$,

$$\boxed{\theta_\kappa(x,\xi) := \frac{1}{2}\left(x\xi - y\eta\right)\Big|_{(y,\eta)=\kappa(x,\xi)},} \tag{3.4.3}$$

and we notice that if κ_1 and κ_2 are two such transformations, then

$$\theta_{\kappa_1\circ\kappa_2}(x,\xi) = \frac{1}{2}\left(x\xi - y\eta\right)\Big|_{(y,\eta)=\kappa_2(x,\xi)} + \frac{1}{2}\left(y\eta - z\zeta\right)\Big|_{\substack{(y,\eta)=\kappa_2(x,\xi)\\(z,\zeta)=\kappa_1(y,\eta)}}$$

and therefore

$$\theta_{\kappa_1\circ\kappa_2} = \theta_{\kappa_2} + \theta_{\kappa_1}\circ\kappa_2. \tag{3.4.4}$$

In the three particular cases above, we get

$$\theta_{j_B} = 0;\ \ \theta_{k_C}(x,\xi) = -\frac{1}{2}\left\langle Cx, x\right\rangle;\ \ \theta_{\ell_j}(x,\xi) = x_j\xi_j. \tag{3.4.5}$$

As we shall see, these functions will appear as phase shifts when we make T_A act on J_B, K_C, or L_j, respectively. Writing $A=(a_{j,k})_{1\le j,k\le n}$ we also define

$$\begin{aligned}
\mathcal{M}_{j_B}(A) &= {}^tB^{-1}AB^{-1},\\
\mathcal{M}_{k_C}(A) &= A + iC,\\
\mathcal{M}_{\ell_j}(A) &= \widetilde{A}_j - \frac{1}{a_{j,j}}\left(a_{j,k}a_{j,l}(1-\delta_{j,k})(1-\delta_{j,l})\right)_{1\le k,l\le n},
\end{aligned}$$

where $\delta_{j,k}$ is the usual Kronecker symbol and $\widetilde{A}_j$ is obtained from A by substituting

$$R_j = \frac{1}{a_{j,j}}(-ia_{j,1},\dots,-ia_{j,j-1},1,-ia_{j,j+1},\dots,-ia_{j,n})$$

into its jth row, and ${}^tR_j$ into its jth column.

In fact, there is a more systematic way to define $\mathcal{M}_\kappa(A)$ (which can be extended for any linear canonical transformation κ), as can be seen from the following result:

Lemma 3.4.1 *For $\kappa\in\{j_B, k_C, \ell_j\}$, one has*

$$\boxed{\kappa\left(\{(x, i\overline{A}x)\ ;\ x\in\mathbf{C}^n\}\right) = \{(y, i\overline{\mathcal{M}_\kappa(A)}y)\ ;\ y\in\mathbf{C}^n\}.} \tag{3.4.6}$$

More generally, for any linear canonical transformation κ on $\mathbf{R}^{2n}$, there exists a unique symmetric matrix $\mathcal{M}_\kappa(A)$ such that $\mathrm{Re}\,\mathcal{M}_\kappa(A)$ is positive definite and (3.4.6) is valid.

Proof If $\kappa \in \{j_B, k_C, \ell_j\}$, the identity (3.4.6) can be verified by a straightforward computation, and is left as an exercise to the reader. Now let us consider the case where κ is a general linear canonical transformation on $\mathbf{R}^{2n}$. For $X = (x, i\overline{A}x)$ with $x \in \mathbf{C}^n$, we have (extending σ on $\mathbf{C}^{2n} \times \mathbf{C}^{2n}$ by $\mathbf{C}$-linearity)

$$\sigma(X, \overline{X}) = iAx \cdot \overline{x} + ix \cdot \overline{A}\overline{x},$$

and therefore, since A is symmetric and $\mathrm{Re}A$ is positive definite,

$$\frac{1}{2i}\sigma\left(X, \overline{X}\right) = (\mathrm{Re}A)x \cdot \overline{x} \geq \frac{1}{C_0}|x|^2 \tag{3.4.7}$$

for some positive constant C_0. Now, extending κ to $\mathbf{C}^{2n}$ by $\mathbf{C}$-linearity, we also have

$$\sigma\left(\kappa(X), \overline{\kappa(X)}\right) = \sigma\left(\kappa(X), \kappa\left(\overline{X}\right)\right) = \sigma\left(X, \overline{X}\right), \tag{3.4.8}$$

where the last equality comes from the fact that κ is canonical. On the other hand, writing

$$(y, \eta) = \kappa(X),$$

one has

$$\sigma\left(\kappa(X), \overline{\kappa(X)}\right) = \eta \cdot \overline{y} - y \cdot \overline{\eta} = 2i\,\mathrm{Im}(\eta \cdot \overline{y}). \tag{3.4.9}$$

In particular, we deduce from (3.4.7)-(3.4.9) that for any $x \in \mathbf{C}^n$, if $(y, \eta) = \kappa(x, i\overline{A}x)$, then

$$|x|^2 \leq C_0 \mathrm{Im}(\eta \cdot \overline{y}). \tag{3.4.10}$$

As a consequence, setting $y = 0$ in (3.4.10) we deduce from it that

$$\kappa\left(\{(x, i\overline{A}x)\ ;\ x \in \mathbf{C}^n\}\right) \cap \{0\} \times \mathbf{C}^n = \{(0,0)\}.$$

This means that the subspace $\kappa\left(\{(x, i\overline{A}x)\ ;\ x \in \mathbf{C}^n\}\right)$ of $\mathbf{C}^{2n}$ is transversal to $\{y = 0\}$, and thus there exists a (unique) matrix $\mathcal{M}_\kappa(A)$ such that

$$\kappa\left(\{(x, i\overline{A}x)\ ;\ x \in \mathbf{C}^n\}\right) = \{(y, i\overline{\mathcal{M}_\kappa(A)}y)\ ;\ y \in \mathbf{C}^n\}.$$

It remains to show that $\mathcal{M}_\kappa(A)$ is symmetric and has a positive definite real part. The fact that it is symmetric is just a consequence of the identity

$$\sigma(\kappa(x, i\overline{A}x), \kappa(x', i\overline{A}x')) = \sigma((x, i\overline{A}x), (x', i\overline{A}x')) = i\overline{A}x.x' - x.i\overline{A}x' = 0$$

for any $x, x' \in \mathbf{C}^n$, which gives

$$0 = \sigma((y, i\overline{\mathcal{M}_\kappa(A)}y), (y', i\overline{\mathcal{M}_\kappa(A)}y')) = i\overline{\mathcal{M}_\kappa(A)}y.y' - y.i\overline{\mathcal{M}_\kappa(A)}y'$$

for any $y, y' \in \mathbf{C}^n$. (More generally, a subspace $\Lambda \subset \mathbf{C}^{2n}$ is said to be *isotropic* if the application $\Lambda^2 \ni (X, Y) \mapsto \sigma(X, Y)$ vanishes identically, and it is clear that such a property is conserved by canonical transformations; if, moreover, $\dim\Lambda = n$, then Λ is said to be *Lagrangian*, and this is again conserved by canonical transformations.)

Finally, using (3.4.10), we get that for any $y \in \mathbf{R}^n\backslash\{0\}$, one has

$$\mathrm{Im}\,(i\overline{\mathcal{M}_\kappa(A)}y \cdot y) > 0,$$

that is,

$$\mathrm{Re}\mathcal{M}_\kappa(A)y \cdot y > 0,$$

which means that $\mathrm{Re}\mathcal{M}_\kappa(A)$ is positive definite. ◇

Remark 3.4.2 If κ_1 and κ_2 are two linear canonical transformations on $\mathbf{R}^{2n}$, Lemma 3.4.1 (applied to $\mathcal{M}_{\kappa_2}(A)$ instead of A) permits us to define the matrix $\mathcal{M}_{\kappa_1}(\mathcal{M}_{\kappa_2}(A))$. Then by construction one also has

$$\mathcal{M}_{\kappa_1}(\mathcal{M}_{\kappa_2}(A)) = \mathcal{M}_{\kappa_1 \circ \kappa_2}(A). \tag{3.4.11}$$

Now we prove the following result:

Proposition 3.4.3 *If κ denotes any one of the previous transformations j_B, k_C, or ℓ_j, denote by $\mathcal{J}_\kappa$ the corresponding operator J_B, K_C, or L_j, respectively. Then for all $u \in C_0^\infty(\mathbf{R}^n)$ one has*

$$T_A \mathcal{J}_\kappa u = \beta e^{i\theta_\kappa/h}\left(T_{\mathcal{M}_\kappa(A)}u\right) \circ \kappa,$$

where $\beta = 1$ if $\kappa \in \{j_B, k_C\}$, and $\beta = \sqrt{\dfrac{|a_{j,j}|}{a_{j,j}}}$ if $\kappa = \ell_j$ (here the determination of the square root on $\mathbf{R}_+^* + i\mathbf{R}$ is the one that assigns positive numbers to positive numbers).

Proof Let us start with the first type. By the change of variables $y \mapsto z = By$, we have

$$
\begin{aligned}
&T_A J_B u(x,\xi)\\
&\quad = \alpha_{n,h}(\det A_1)^{\frac14}|\det B|^{\frac12}\int e^{i(x-y)\xi/h-q_A(x-y)/2h}u(By)dy\\
&\quad = \alpha_{n,h}(\det A_1)^{\frac14}|\det B|^{-\frac12}\int e^{i(x-B^{-1}z)\xi/h-q_A(x-B^{-1}z)/2h}u(z)dz\\
&\quad = \alpha_{n,h}(\det{}^tB^{-1}A_1B^{-1})^{\frac14}\int e^{i(Bx-z)^tB^{-1}\xi/h-q_{{}^tB^{-1}AB^{-1}}(Bx-z)/2h}u(z)dz\\
&\quad = T_{{}^tB^{-1}AB^{-1}}u(Bx,{}^tB^{-1}\xi).
\end{aligned}
$$

For the second type, we use the fact that

$$
q_A(x-y)+i\langle Cy,y\rangle = q_{A+iC}(x-y)-i\langle Cx,x\rangle+2i\langle Cx,y\rangle,
$$

which gives

$$
\begin{aligned}
&T_A K_C u(x,\xi)\\
&\quad = \alpha_{n,h}(\det A_1)^{\frac14}\int e^{i(x-y)\xi/h+i\langle Cx,x\rangle/2h-i\langle Cx,y\rangle/h-q_{A+iC}(x-y)/2h}u(y)dy\\
&\quad = \alpha_{n,h}(\det A_1)^{\frac14}e^{-i\langle Cx,x\rangle/2h}\int e^{i(x-y)(\xi+Cx)/h-q_{A+iC}(x-y)/2h}u(y)dy\\
&\quad = e^{-i\langle Cx,x\rangle/2h}T_{A+iC}u(x,\xi+Cx).
\end{aligned}
$$

For the third type, by a permutation of the variables (and an application of the result for the first type), we can assume that $j=1$. Then denoting $y'=(y_2,\dots,y_n)\in\mathbf{R}^{n-1}$ for $y=(y_1,y_2,\dots,y_n)\in\mathbf{R}^n$, we have

$$
\begin{aligned}
&T_A L_1 u(x,\xi)\\
&\quad = \frac{\alpha_{n,h}(\det A_1)^{\frac14}}{\sqrt{2\pi h}}\int e^{i(x-y)\xi/h-q_A(x-y)/2h-iy_1z_1/h}u(z_1,y')dz_1\,dy\\
&\quad = \frac{\alpha_{n,h}(\det A_1)^{\frac14}}{\sqrt{2\pi h}}\int e^{[i(x_1-y_1)(\xi_1+z_1)-q_A(x-y)/2-ix_1z_1+i(x'-y')\xi']/h}u(z_1,y')dz_1\,dy\\
&\quad = \frac{\alpha_{n,h}(\det A_1)^{\frac14}}{\sqrt{2\pi h}}\int e^{iy_1(\xi_1+z_1)/h-q_A(y)/2h-ix_1z_1/h+iy'\xi'/h}u(z_1,x'-y')dz_1\,dy,
\end{aligned}
$$

where the last equality comes from the change of variables: $y\mapsto x-y$. Now, $q_A(y)=a_{1,1}y_1^2+2\sum_{j\geq 2}a_{1,j}y_1y_j+\sum_{j,k\geq 2}a_{j,k}y_jy_k$ and since $\mathrm{Re}\,a_{1,1}>0$, we have

$$
\int e^{iy_1(\xi_1+z_1)/h-a_{1,1}y_1^2/2h-\sum_{j\geq 2}a_{1,j}y_1y_j/h}dy_1=\sqrt{\frac{2\pi h}{a_{1,1}}}e^{-(\xi_1+z_1+i\sum_{j\geq 2}a_{1,j}y_j)^2/2a_{1,1}h}.
$$

Therefore,

$$
\begin{aligned}
&T_A L_1 u(x,\xi)\\
&\quad = \frac{\alpha_{n,h}(\det A_1)^{\frac14}}{\sqrt{a_{1,1}}}\int e^{[-(\xi_1+z_1+i\sum_{j\geq 2}a_{1,j}y_j)^2/a_{1,1}-\sum_{j,k\geq 2}a_{j,k}y_jy_k-2ix_1z_1+2iy'\xi']/2h}\\
&\qquad\qquad \times u(z_1,x'-y')dz_1\,dy'. \qquad (3.4.12)
\end{aligned}
$$

On the other hand, by the same change of variables we have

$$
\begin{aligned}
&e^{ix_1\xi_1/h}T_{\mathcal{M}_{\ell_1}(A)}u(-\xi_1,x';x_1,\xi')\\
&\quad = \alpha_{n,h}\left(\det \operatorname{Re}\mathcal{M}_{\ell_1}(A)\right)^{\frac14}\int e^{-ix_1z_1/h+iy'\xi'/h-q_{\mathcal{M}_{\ell_1}(A)}(-\xi_1-z_1,y')/2h}\\
&\qquad\qquad \times u(z_1,x'-y')dz_1\,dy', \qquad (3.4.13)
\end{aligned}
$$

and using the definition of $\mathcal{M}_{\ell_1}(A)$, we see that

$$
\begin{aligned}
&q_{\mathcal{M}_{\ell_1}(A)}(-\xi_1-z_1,y')\\
&\quad = \frac{1}{a_{1,1}}(\xi_1+z_1)^2+2i\sum_{j\geq 2}\frac{a_{1,j}}{a_{1,1}}(\xi_1+z_1)y_j+\sum_{j,k\geq 2}\left(a_{j,k}-\frac{a_{1,j}a_{1,k}}{a_{1,1}}\right)y_jy_k\\
&\quad = \left(\xi_1+z_1+i\sum_{j\geq 2}a_{1,j}y_j\right)^2/a_{1,1}+\sum_{j,k\geq 2}a_{j,k}y_jy_k. \qquad (3.4.14)
\end{aligned}
$$

We deduce from (3.4.12)-(3.4.14) that there exists a complex constant γ such that

$$T_AL_1u(x,\xi)=\gamma e^{ix_1\xi_1/h}T_{\mathcal{M}_{\ell_1}(A)}u(-\xi_1,x';x_1,\xi'). \qquad (3.4.15)$$

Moreover, since both $T_A\circ L_1$ and $T_{\mathcal{M}_{\ell_1}(A)}$ are isometries from $L^2(\mathbf{R}^n)$ to $L^2(\mathbf{R}^{2n})$, we have necessarily

$$|\gamma|=1, \qquad (3.4.16)$$

and applying (3.4.15) to $u=\delta$ (the Dirac measure at 0) and $(x,\xi)=(0,0)$, we get in particular (using also (3.4.12))

$$\frac{(\det A_1)^{\frac14}}{\sqrt{a_{1,1}}}=\gamma\left(\det \operatorname{Re}\mathcal{M}_{\ell_1}(A)\right)^{\frac14}. \qquad (3.4.17)$$

As a consequence, $(\gamma\sqrt{a_{1,1}}) \in \mathbf{R}_+^*$, and thus by (3.4.16),

$$\gamma = \sqrt{\frac{|a_{1,1}|}{a_{1,1}}}. \tag{3.4.18}$$

In view of (3.4.15), this finishes the proof of Proposition 3.4.3. ◇

Remark 3.4.4 As a particular case, we get

$$\boxed{T\mathcal{F}_h u(x,\xi) = e^{ix\xi/h} Tu(-\xi, x).}$$

Remark 3.4.5 Incidentally, we have also proved

$$\det \mathrm{Re}\mathcal{M}_{\ell_j}(A) = \frac{1}{|a_{j,j}|^2} \det \mathrm{Re} A.$$

Remark 3.4.6 If κ_1 and κ_2 are two transformations of the type j_B, k_C, or ℓ_j, then applying Proposition 3.4.3 twice we get

$$\begin{aligned} T_A \mathcal{J}_{\kappa_2} \mathcal{J}_{\kappa_1} u &= \beta_2 e^{i\theta_{\kappa_2}/h} \left(T_{\mathcal{M}_{\kappa_2}(A)} \mathcal{J}_{\kappa_1} u \right) \circ \kappa_2 \\ &= \beta_2 e^{i\theta_{\kappa_2}/h} \left(\beta_1 e^{i\theta_{\kappa_1}/h} T_{\mathcal{M}_{\kappa_1}(\mathcal{M}_{\kappa_2}(A))} u \circ \kappa_1 \right) \circ \kappa_2 \end{aligned}$$

(with $|\beta_1| = |\beta_2| = 1$) and therefore, using (3.4.4) and (3.4.11),

$$T_A \mathcal{J}_{\kappa_2} \mathcal{J}_{\kappa_1} u = \beta_1 \beta_2 e^{i\theta_{\kappa_1 \circ \kappa_2}/h} T_{\mathcal{M}_{\kappa_1 \circ \kappa_2}(A)} u \circ (\kappa_1 \circ \kappa_2). \tag{3.4.19}$$

Now, as we have already noticed, if κ is a general linear canonical transformation on $\mathbf{R}^{2n}$, one can show that it can always be written in the form (see, e.g., [Fo], Chapter 4, for a proof of this fact)

$$\kappa = \kappa_1 \circ \kappa_2 \circ \ldots \circ \kappa_N \tag{3.4.20}$$

for some $N \in \mathbf{N}$, where for any $\nu \in \{1, \ldots, N\}$, κ_ν belongs to one of the previous forms j_B, k_C, or ℓ_j. Of course, there is no unicity in the way of writing κ as in (3.4.20), but given such an expression, if we set

$$\mathcal{J}_\kappa = \mathcal{J}_{\kappa_N} \mathcal{J}_{\kappa_{N-1}} \cdots \mathcal{J}_{\kappa_1}, \tag{3.4.21}$$

then an iteration of (3.4.19) shows that for any $u \in C_0^\infty(\mathbf{R}^n)$, one has

$$\boxed{T_A \mathcal{J}_\kappa u = \beta_\kappa e^{i\theta_\kappa/h} \left(T_{\mathcal{M}_\kappa(A)} u \right) \circ \kappa,} \tag{3.4.22}$$

where β_κ is an h-independent complex constant of modulus 1.

3.5 Microlocal Exponential Estimates

In view of studying MS(u) for u a solution of a partial differential equation of the type

$$P(x, hD_x)u = 0$$

with $P(x, \xi)$ analytic, we first establish some a priori estimates involving Tu.

As has been proved in Proposition 3.3.1, any pseudodifferential operator on $\mathbf{R}^n$ is transformed by T into a pseudodifferential operator on $\mathbf{R}^{2n}$. Moreover, multiplying P by a convenient elliptic pseudodifferential operator, one can reduce to the case of a *bounded* pseudodifferential operator. For these reasons, we start by considering the case of a bounded pseudodifferential operator on $\mathbf{R}^{2n}$:

$$Q = \mathrm{Op}_t(q(x, \xi, x^*, \xi^*)),$$

where $q \in S_{4n}(1)$ and $t \in [0, 1]$. Let also $\psi = \psi(x, \xi) \in S_{2n}(1)$ be a *real-valued* smooth function on $\mathbf{R}^{2n}$. Then we have the following theorem:

Theorem 3.5.1 *There exist $\tilde{q}(x, \xi; h) \in S_{2n}(1)$ and $R(h) \in \mathcal{L}(L^2(\mathbf{R}^{2n}))$ such that for all $u, v \in L^2(\mathbf{R}^n)$, one has*

$$\left\langle Qe^{\psi/h}Tu, e^{\psi/h}Tv\right\rangle_{L^2(\mathbf{R}^{2n})} = \left\langle(\tilde{q}(x, \xi; h) + R(h))e^{\psi/h}Tu, e^{\psi/h}Tv\right\rangle_{L^2(\mathbf{R}^{2n})}$$

and

$$\begin{cases} \tilde{q}(x, \xi; h) \sim \sum_{j\geq 0} h^j \tilde{q}_j(x, \xi) \text{ in } S_{2n}(1), \\ \tilde{q}_0(x, \xi) = q(x, \xi, \xi - \partial_\xi\psi(x, \xi), \partial_x\psi(x, \xi)), \\ \|R(h)\|_{\mathcal{L}(L^2(\mathbf{R}^{2n}))} = \mathcal{O}(h^\infty), \end{cases}$$

uniformly as $h \to 0_+$.

Remark 3.5.2 In fact, by an argument of density it will follow from the proof that this formula can be extended to those $\psi \in C^\infty(\mathbf{R}^{2n}\,; \mathbf{R})$ such that $\nabla\psi \in S_{2n}(1)$ (ψ not necessarily bounded), on the condition that u and v belong to the space $\mathcal{H}_\psi$ defined as the completion of $C_0^\infty(\mathbf{R}^n)$ in the norm $\|u\|_\psi := \|e^{\psi/h}Tu\|_{L^2}$.

Proof of Theorem 3.5.1 The proof we present here is essentially taken from [Na2]. Let

$$r_1(x, \xi, x^*, \xi^*) = q(x, \xi, x^*, \xi^*) - q(x, \xi, \xi - \partial_\xi\psi, \partial_x\psi).$$

Then since r_1 vanishes on $\{x^* - \xi + \partial_\xi\psi = \xi^* - \partial_x\psi = 0\}$, by Taylor's formula there exist two (vector-valued) smooth functions $q_1 = q_1(x, \xi, x^*, \xi^*)$ and $q_2 = q_2(x, \xi, x^*, \xi^*)$ such that

$$r_1 = (x^* - \xi + \partial_\xi\psi)q_1 + (\xi^* - \partial_x\psi)q_2. \tag{3.5.1}$$

In fact, we have

$$\begin{aligned} q_1 &= \int_0^1 (\partial_{x^*} q)(x, \xi, \xi - \partial_\xi\psi + t(x^* - \xi + \partial_\xi\psi), \xi^*)\,dt, \\ q_2 &= \int_0^1 (\partial_{\xi^*} q)(x, \xi, \xi - \partial_\xi\psi, \partial_x\psi + t(\xi^* - \partial_x\psi))\,dt, \end{aligned}$$

and thus $q_1, q_2 \in S_{4n}(1)$. Set

$$\begin{aligned} F &= hD_x - \xi + \partial_\xi\psi, \\ G &= hD_\xi - \partial_x\psi, \\ Q_j &= \mathrm{Op}_h^t(q_j) \quad (j = 1, 2). \end{aligned} \tag{3.5.2}$$

Then, using the symbolic calculus of pseudodifferential operators (see Section 2.7), we can deduce from (3.5.1) that there exists $r_2 \in S_{4n}(1)$ such that

$$\mathrm{Op}_h^t(r_1) = \frac{1}{2}(Q_1F + FQ_1) + \frac{1}{2}(Q_2G + GQ_2) + h\mathrm{Op}_h^t(r_2).$$

Moreover, we have seen in Proposition 3.1.1 that

$$(hD_x - \xi)T = ihD_\xi T,$$

and therefore, setting

$$T_\psi : u \mapsto e^{\psi/h}Tu$$

we also have

$$(hD_x - \xi + i\partial_x\psi)T_\psi = (ihD_\xi - \partial_\xi, \psi)T_\psi$$

that is,

$$\boxed{FT_\psi = iGT_\psi.} \tag{3.5.3}$$

It is this last equality that will permit to us to conclude, and let us notice that its main originality relies on the fact that it identifies the action of the *symmetric* operator F with that of the *antisymmetric* operator iG in the range of T.

Now, for all $u, v \in C_0^\infty(\mathbf{R}^n)$, we have

$$\begin{aligned}
&\left\langle \mathrm{Op}_h^t(r_1)T_\psi u, T_\psi v\right\rangle \\
&\quad = \frac{1}{2}\left(\langle (Q_1F + FQ_1)T_\psi u, T_\psi v\rangle + \langle (Q_2G + GQ_2)T_\psi u, T_\psi v\rangle\right) \qquad (3.5.4)\\
&\qquad + h\left\langle \mathrm{Op}_h^t(r_2)T_\psi u, T_\psi v\right\rangle,
\end{aligned}$$

and by (3.5.3),

$$\begin{aligned}
\langle FQ_1T_\psi u, T_\psi v\rangle &= \langle Q_1T_\psi u, FT_\psi v\rangle\\
&= \langle Q_1T_\psi u, iGT_\psi v\rangle\\
&= -i\langle GQ_1T_\psi u, T_\psi v\rangle\\
&= -i\langle Q_1GT_\psi u, T_\psi v\rangle + i\langle [Q_1, G]T_\psi u, T_\psi v\rangle\\
&= -\langle Q_1FT_\psi u, T_\psi v\rangle + i\langle [Q_1, G]T_\psi u, T_\psi v\rangle,
\end{aligned}$$

which gives

$$\frac{1}{2}\langle (FQ_1 + Q_1F)T_\psi u, T_\psi v\rangle = \frac{i}{2}\langle [Q_1, G]T_\psi u, T_\psi v\rangle. \qquad (3.5.5)$$

In a similar way, we also have

$$\frac{1}{2}\langle (Q_2G + GQ_2)T_\psi u, T_\psi v\rangle = \frac{i}{2}\langle [F, Q_2]T_\psi u, T_\psi v\rangle \qquad (3.5.6)$$

and therefore, substituting in (3.5.4), we get

$$\left\langle \mathrm{Op}_h^t(r_1)T_\psi u, T_\psi v\right\rangle = \left\langle \left(\frac{i}{2}[Q_1, G] + \frac{i}{2}[F, Q_2] + h\mathrm{Op}_h^t(r_2)\right)T_\psi u, T_\psi v\right\rangle. \qquad (3.5.7)$$

Now we observe (still using the symbolic calculus of Section 2.7) that

$$\frac{i}{2}[Q_1, G] + \frac{i}{2}[F, Q_2] + h\mathrm{Op}_h^t(r_2) = h\mathrm{Op}_h^t(q')$$

for some $q' \in S_{4n}(1)$, and thus

$$\left\langle \mathrm{Op}_h^t(r_1)T_\psi u, T_\psi v\right\rangle = h\left\langle \mathrm{Op}_h^t(q')T_\psi u, T_\psi v\right\rangle. \qquad (3.5.8)$$

Summing up, until now we have proved that for any $q \in S_{4n}(1)$, there exists $q' \in S_{4n}(1)$ such that

$$\left\langle \mathrm{Op}_h^t(q) T_\psi u, T_\psi v \right\rangle = \langle q(x,\xi,\xi - \partial_\xi \psi, \partial_x \psi) T_\psi u, T_\psi v \rangle + h \left\langle \mathrm{Op}_h^t(q') T_\psi u, T_\psi v \right\rangle . \tag{3.5.9}$$

Performing the same argument for q', and iterating the procedure, we get that there exist a sequence $(\tilde{q}_j)_{j\in\mathbf{N}}$ of elements of $S_{2n}(1)$, and a sequence $(q^{(j)})_{j\in\mathbf{N}}$ of elements of $S_{4n}(1)$, such that at any order $N \in \mathbf{N}$ one has

$$\left\langle \mathrm{Op}_h^t(q) T_\psi u, T_\psi v \right\rangle = \left\langle \sum_{j=0}^{N-1} h^j \tilde{q}_j T_\psi u, T_\psi v \right\rangle + h^N \left\langle \mathrm{Op}_h^t(q^{(N)}) T_\psi u, T_\psi v \right\rangle \tag{3.5.10}$$

with $\tilde{q}_0(x,\xi) = q(x,\xi,\xi - \partial_\xi \psi, \partial_x \psi)$.

Now let $\tilde{q} \in S_{2n}(1)$ be a resummation of $\sum_{j\geq 0} h^j \tilde{q}_j$ in the sense of Proposition 2.3.2, and write

$$R = \Pi_\psi \left(\mathrm{Op}_h^t(q) - \tilde{q} \right) \Pi_\psi, \tag{3.5.11}$$

where Π_ψ denotes the orthogonal projection from $L^2(\mathbf{R}^{2n})$ onto the image of T_ψ (which is closed in $L^2(\mathbf{R}^{2n})$, since the image of T is isometric to $L^2(\mathbf{R}^n)$ by Proposition 3.1.1, and the map $w \mapsto e^{\psi/h} w$ is closed on $L^2(\mathbf{R}^{2n})$). By (3.5.10) and the Calderón–Vaillancourt theorem, we see that for any $N \in \mathbf{N}$, one has

$$\langle R T_\psi u, T_\psi v \rangle = \mathcal{O}(h^N) \|T_\psi u\| \cdot \|T_\psi v\|$$

and thus, since R leaves the image of T_ψ stable and vanishes on its orthogonal:

$$\|R\|_{\mathcal{L}(L^2(\mathbf{R}^{2n}))} = \mathcal{O}(h^\infty). \tag{3.5.12}$$

In view of the definition (3.5.11) of R, the estimate (3.5.12) gives exactly the required result, and this finishes the proof of Theorem 3.5.1. ◇

Now let $a, b > 0$ and $p \in S_{2n}(1)$ such that p extends holomorphically to the complex strip

$$\Sigma(a,b) := \{(x,\xi) \in \mathbf{C}^{2n} \; ; \; |\mathrm{Im} x| < a \; , \; |\mathrm{Im} \xi| < b\}$$

and satisfies

$$\forall\, \alpha \in \mathbf{N}^{2n} \; , \; \partial^\alpha p = \mathcal{O}(1) \text{ uniformly in } \Sigma(a,b). \tag{3.5.13}$$

Assume also that the real-valued function $\psi \in S_{2n}(1)$ satisfies

$$\sup_{\mathbf{R}^{2n}} |\nabla_x \psi| < b \qquad \sup_{\mathbf{R}^{2n}} |\nabla_\xi \psi| < a, \tag{3.5.14}$$

and for $t \in [0,1]$ fixed set

$$P = \mathrm{Op}_h^t(p).$$

Then we have the following important corollary of Theorem 3.5.1:

Corollary 3.5.3 *Let $f \in S_{2n}(1)$. Under assumptions (3.5.13)-(3.5.14), there exist $\tilde{p}(x,\xi;h) \in S_{2n}(1)$ and $R(h) \in \mathcal{L}(L^2(\mathbf{R}^{2n}))$ such that for all $u, v \in L^2(\mathbf{R}^n)$, one has*

$$\left\langle f e^{\psi/h} TPu, e^{\psi/h} Tv \right\rangle_{L^2(\mathbf{R}^{2n})} = \left\langle (\tilde{p}(x,\xi;h) + R(h)) e^{\psi/h} Tu, e^{\psi/h} Tv \right\rangle_{L^2(\mathbf{R}^{2n})}$$

and

$$\begin{cases} \tilde{p}(x,\xi;h) \sim \sum_{j\geq 0} h^j \tilde{p}_j(x,\xi) \text{ in } S_{2n}(1), \\ \tilde{p}_0(x,\xi) = f(x,\xi) p(x - 2\partial_z \psi(x,\xi), \xi + 2i\partial_z \psi(x,\xi)), \\ \|R(h)\|_{\mathcal{L}(L^2(\mathbf{R}^{2n}))} = \mathcal{O}(h^\infty), \end{cases}$$

where

$$\partial_z := \frac{1}{2}(\nabla_x + i\nabla_\xi)$$

is holomorphic differentiation with respect to $z = x - i\xi$.

Proof By Proposition 3.3.1, we have

$$f e^{\psi/h} TPu = Q e^{\psi/h} Tu$$

with

$$Q = f e^{\psi/h} \mathrm{Op}_h^t(p(x - \xi^*, x^*)) e^{-\psi/h}. \tag{3.5.15}$$

Therefore, in view of applying Theorem 3.5.1 we first have to prove the following lemma:

Lemma 3.5.4 *The operator $f e^{\psi/h} \mathrm{Op}_h^t(p(x - \xi^*, x^*)) e^{-\psi/h}$ is a semiclassical pseudodifferential operator on $\mathbf{R}^{2n}$, and its symbol $q \in S_{4n}(1)$ admits an asymptotic expansion of the form*

$$q \sim \sum_{j\geq 1} h^j q_j \text{ in } S_{4n}(1)$$

with

$$q_0(x,\xi,x^*,\xi^*) = f(x,\xi) p(x - \xi^* - i\partial_\xi \psi, x^* + i\partial_x \psi).$$

Proof Here again we follow [Na2]. Set $\widetilde{P} = \mathrm{Op}_h^t(p(x-\xi^*, x^*))$. For $w \in C_0^\infty(\mathbf{R}^{2n})$ and $\nu \in \mathbf{R}$, we write

$$e^{i\nu\psi/h}\widetilde{P}e^{-i\nu\psi/h}w(x,\xi)$$

$$= \frac{1}{(2\pi h)^{2n}}\int e^{i(x-y)x^*/h+i(\xi-\eta)\xi^*/h+i\nu(\psi(x,\xi)-\psi(y,\eta))/h}$$
$$\times p((1-t)x+ty-\xi^*, x^*))w(y,\eta)dy\,d\eta\,dx^*\,d\xi^*,$$

and we also have by Taylor's formula

$$\psi(x,\xi)-\psi(y,\eta) = (x-y)\phi_1(x,y,\xi,\eta)+(\xi-\eta)\phi_2(x,y,\xi,\eta),$$

where $\phi_1, \phi_2 \in [S_{4n}(1)]^n$ are real-valued.

Then we make the change of variables $(x^*, \xi^*) \mapsto (\widetilde{x}^*, \widetilde{\xi}^*)$ given by

$$\begin{cases} \widetilde{x}^* = x^* + \nu\phi_1(x,y,\xi,\eta), \\ \widetilde{\xi}^* = \xi^* + \nu\phi_2(x,y,\xi,\eta). \end{cases} \tag{3.5.16}$$

We obtain

$$e^{i\nu\psi/h}\widetilde{P}e^{-i\nu\psi/h}w(x,\xi)$$

$$= \frac{1}{(2\pi h)^{2n}}\int e^{i(x-y)\widetilde{x}^*/h+i(\xi-\eta)\widetilde{\xi}^*/h} \tag{3.5.17}$$
$$\times p((1-t)x+ty-\widetilde{\xi}^*+\nu\phi_2, \widetilde{x}^*-\nu\phi_1)w(y,\eta)dy\,d\eta\,d\widetilde{x}^*\,d\widetilde{\xi}^*,$$

and since $|\phi_1| \leq \sup|\nabla_x\psi| < b$ and $|\phi_2| \leq \sup|\nabla_\xi\psi| < a$, we see that the function

$$\nu \mapsto p((1-t)x+ty-\widetilde{\xi}^*+\nu\phi_2, \widetilde{x}^*-\nu\phi_1)$$

can be extended holomorphically in a complex neighborhood of $\{\nu \in \mathbf{C}\,;\, |\nu| \leq 1\}$, with values in $S_{6n}(1)$. As a consequence, the right-hand side of (3.5.17) can be extended too, and since this is also obviously true for the left-hand side and both sides are equal for $\nu \in \mathbf{R}$, by analytic continuation they remain equal for $\nu \in \mathbf{C}$, $|\nu| \leq 1$. In particular, for $\nu = -i$ we get

$$e^{\psi/h}\widetilde{P}e^{-\psi/h} = \mathrm{Op}(p((1-t)x+ty-\xi^*-i\phi_2, x^*+i\phi_1)),$$

where the quantization is given by the general case of symbols in $S_{3d}(1)$ (with $d = 2n$ here) as in Definition 2.5.1. Then, using Theorem 2.7.1, we obtain

$$e^{\psi/h}\widetilde{P}e^{-\psi/h} = \mathrm{Op}_h^t(p_t(x,\xi,x^*,\xi^*;\ h)), \tag{3.5.18}$$

where p_t admits an asymptotic expansion in $S_{4n}(1)$, with first term given by

$$\begin{aligned} p_t^0(x,\xi,x^*,\xi^*) &= p((1-t)x + ty - \xi^* - i\phi_2, x^* + i\phi_1)\Big|_{\substack{y=x\\ \eta=\xi}} \\ &= p(x - \xi^* - i\partial_\xi\psi, x^* + i\partial_x\psi). \end{aligned} \tag{3.5.19}$$

Therefore, the lemma follows from (3.5.18) and (3.5.19) by multiplying (3.5.18) by $f(x,\xi)$. ◇

End of the proof of Corollary 3.5.3 From Lemma 3.5.4, we see that we can apply Theorem 3.5.1 with Q given in (3.5.15), and this gives exactly the result of Corollary 3.5.3. ◇

In the same situation as for Corollary 3.5.3 there is another consequence of Theorem 3.5.1 that will be rather useful in the applications:

Corollary 3.5.5 *Let $f \in S_{2n}(1)$. Under assumptions (3.5.13)-(3.5.14) one has*

$$\|fe^{\psi/h}TPu\|^2 = \|f(x,\xi)p(x - 2\partial_z\psi, \xi + 2i\partial_z\psi)e^{\psi/h}Tu\|^2 + \mathcal{O}(h)\|e^{\psi/h}Tu\|^2$$

uniformly with respect to $u \in L^2(\mathbf{R}^n)$ and $h > 0$ small enough.

Proof : With Q given in (3.5.15), one has

$$\|fe^{\psi/h}TPu\|^2 = \left\langle Q^*Qe^{\psi/h}Tu, e^{\psi/h}Tu\right\rangle$$

and from Lemma 3.5.4 and the symbolic calculus, Q^*Q is a semiclassical pseudodifferential operator whose symbol admits an asymptotic expansion with first term $|f(x,\xi)p(x - \xi^* - i\partial_\xi\psi, x^* + i\partial_x\psi)|^2$. Then the result follows directly from Theorem 3.5.1. ◇

Remark 3.5.6 Analogous estimates are valid when p is not analytic, but, e.g., Gevrey. In this case, the weight $e^{\psi/h}$ has to be replaced by $e^{\psi/h^{1/s}}$, where $s > 1$ is the Gevrey index, see [Ju]. The proof relies essentially on the almost analytic extensions introduced by Melin and Sjöstrand in [MeSj], and can also be adapted in the general C^∞ case with weights of the type $h^{\psi(x,\xi)}$ (see Exercise 4 of this chapter).

Of course, in the particular case where ψ vanishes identically, one can see from the proof of Corollary 3.5.3 that the assumption of analyticity made on p is no longer necessary. As a consequence, using also Corollary 3.1.4, we get the following result:

Corollary 3.5.7 *Let $p \in S_{2n}(1)$, $t \in [0,1]$, and let $P = \mathrm{Op}_h^t(p)$. Then there exist $\widetilde{p}(x,\xi;h) \in S_{2n}(1)$ and $R(h) \in \mathcal{L}(L^2(\mathbf{R}^{2n}))$ such that for all $u, v \in L^2(\mathbf{R}^n)$, one has*

$$\langle Pu, v\rangle_{L^2(\mathbf{R}^n)} = \langle(\widetilde{p}(x,\xi;h) + R(h))Tu, Tv\rangle_{L^2(\mathbf{R}^{2n})}$$

and

$$\begin{cases} \widetilde{p}(x,\xi;h) \sim \sum_{j\geq 0} h^j \widetilde{p}_j(x,\xi) \text{ in } S_{2n}(1), \\ \widetilde{p}_0(x,\xi) = p(x,\xi), \\ \|R(h)\|_{\mathcal{L}(L^2(\mathbf{R}^{2n}))} = \mathcal{O}(h^\infty). \end{cases}$$

As an application of this last result we have the following semiclassical version of a celebrated theorem (see also Exercise 22 of Chapter 2):

Theorem 3.5.8 (Sharp Gårding Inequality) *Let $p = p(x,\xi) \in S_{2n}(1)$ such that $p \geq 0$ on $\mathbf{R}^{2n}$. Then there exists a constant $C > 0$ such that for all $u \in L^2(\mathbf{R}^n)$ and $h > 0$ small enough, one has*

$$\left\langle \mathrm{Op}_h^W(p)u, u\right\rangle \geq -Ch\|u\|^2.$$

Remark 3.5.9 Since p is real-valued, the operator $\mathrm{Op}_h^W(p)$ is symmetric on $L^2(\mathbf{R}^n)$, and therefore the quantity $\left\langle \mathrm{Op}_h^W(p)u, u\right\rangle$ is necessarily real.

Remark 3.5.10 In other words, the fact that p is nonnegative everywhere implies that (in the sense of operators) $\mathrm{Op}_h^W(p)$ is nonnegative modulo $\mathcal{O}(h)$.

Remark 3.5.11 Of course, the result remains valid for any perturbation of p of order $\mathcal{O}(h)$ in $S_{2n}(1)$. In particular, the assumption $p \geq 0$ can be replaced by $p \geq -C'h$ for some constant C'. Indeed, a stronger result exists for $\mathrm{Op}_h^W(p)$ when $p \geq 0$: This is the so-called *Fefferman–Phong Inequality*, which asserts that in this situation one has

$$\left\langle \mathrm{Op}_h^W(p)u, u\right\rangle \geq -Ch^2\|u\|^2.$$

But the proof is very delicate, and we refer the interested reader to [FePh1] (see also [Bo, Tat]).

Proof of Theorem 3.5.8 It is an immediate consequence of Corollary 3.5.7 and the Calderón–Vaillancourt theorem. $\diamond$

There also exists a generalization of Theorem 3.5.8 to possibly unbounded pseudodifferential operators:

Corollary 3.5.12 *Let $m \in \mathbf{R}$, and $p \in S_{2n}(\langle\xi\rangle^m)$ such that $p \geq 0$ on $\mathbf{R}^{2n}$. Then there exists a constant $C > 0$ such that for all $u \in H^{m/2}(\mathbf{R}^n)$ one has*

$$\left\langle \mathrm{Op}_h^W(p)u, u \right\rangle \geq -Ch\|u\|_{H^{m/2}}^2.$$

Proof By the symbolic calculus, there exists $r \in S_{2n}(1)$ such that

$$\mathrm{Op}_h^W(\langle\xi\rangle^{-m/2})\mathrm{Op}_h^W(p)\mathrm{Op}_h^W(\langle\xi\rangle^{-m/2}) = \mathrm{Op}_h^W(\langle\xi\rangle^{-m}p) + h\mathrm{Op}_h^W(r). \quad (3.5.20)$$

Then the result for $u \in H^{m/2}(\mathbf{R}^n)$ follows from the Calderón–Vaillancourt theorem by applying Theorem 3.5.8 with the symbol $\langle\xi\rangle^{-m}p \in S_{2n}(1)$, and the function $v = \mathrm{Op}_h^W(\langle\xi\rangle^{m/2})u \in L^2(\mathbf{R}^n)$. $\diamond$

Remark 3.5.13 If, moreover, p satisfies $\partial^\alpha p = \mathcal{O}(1+p)$ for all $\alpha \in \mathbf{R}^{2n}$, then the previous inequality can be improved to

$$\left\langle \mathrm{Op}_h^W(p)u, u \right\rangle \geq -Ch\|u\|_{L^2}^2.$$

Actually, in this case a slight generalization of Corollary 3.5.7 gives

$$\langle Pu, u\rangle_{L^2(\mathbf{R}^n)} = \langle (p(x,\xi) + R(h))Tu, Tu\rangle_{L^2(\mathbf{R}^{2n})}$$

with $|\langle R(h)Tu, Tu\rangle| = \mathcal{O}(h\|u\|^2 + h\langle pTu, Tu\rangle)$. Thus the result follows by taking h small enough.

Many other generalizations of Corollary 3.5.3 can be made, including a version whose framework is the so-called *Weyl–Hörmander calculus*, and which should contain all the previous cases. With the notation of [Ho2], and in the case where $f = 1$ and $\psi = 0$, it reads as follows:

For any $a \in S(m, g)$,

$$\left\langle \mathrm{Op}_1^W(a)u, v \right\rangle = \langle (\tilde{a} + R)Tu, Tv\rangle,$$

where T is now defined without the semiclassical parameter, $\tilde{a} \in S(m, g)$ is a symbol with an expansion that can be explicitly computed, and R satisfies, for any $N > 0$,

$$\|Rw\| = \mathcal{O}\left(\left\|h^N w\right\|\right)$$

uniformly with respect to $w \in L^2(\mathbf{R}^{2n})$. Here h is related to the metric g by the formula $h^2 = g/g^\sigma$ (see [Ho2] for the definition of g^σ).

As a final remark of this section, let us note that when ψ depends only on a group of variables, say $(x_{i_1}, \dots, x_{i_k}, \xi_{j_1}, \dots, \xi_{j_\ell})$, then the analyticity assumption in Corollary 3.5.3 can be relaxed with respect to the variables other than $(x_{j_1}, \dots, x_{j_\ell}, \xi_{i_1}, \dots, \xi_{i_k})$. In particular, if $\psi = \psi(x)$ depends only on x (respectively $\psi = \psi(\xi)$ depends only on ξ), then one needs only the analyticity of $p(x, \xi)$ with respect to the variable ξ (respectively x).

3.6 Exercises and Problems

1. **Coherent States** - For $(x, \xi) \in \mathbf{R}^{2n}$ denote by $\phi_{x,\xi}$ the function on $\mathbf{R}^n$ defined by

$$\phi_{x,\xi}(y) = (\pi h)^{-n/4} e^{i(y-x)\xi/h-(y-x)^2/2h}$$

(the so-called *coherent state* centered at (x, ξ)).

(i) Prove that for all $(x, \xi), (x', \xi') \in \mathbf{R}^{2n}$ one has

$$\langle \phi_{x,\xi}, \phi_{x',\xi'} \rangle_{L^2(\mathbf{R}^n)} = e^{i(x'-x)(\xi'+\xi)/2h-(x-x')^2/4h-(\xi-\xi')^2/4h}$$

(in particular, $\|\phi_{x,\xi}\|_{L^2(\mathbf{R}^n)} = 1$).

(ii) Deduce from (i) that for any $(x, \xi) \in \mathbf{R}^{2n}$ one has

$$\mathrm{MS}(\phi_{x,\xi}) = \{(x, \xi)\}.$$

(iii) Use Proposition 3.1.6 to prove that for all $u \in \mathcal{S}'(\mathbf{R}^n)$ one has

$$u = (2\pi h)^{-n/2} \int_{\mathbf{R}^{2n}} Tu(x, \xi) \phi_{x,\xi} dx d\xi$$

(which means that u can be written as a superposition of coherent states).

(iv) If A is a bounded operator on $L^2(\mathbf{R}^n)$ with distribution kernel K_A, then prove the formula

$$K_A(x,y) = (2\pi h)^{-n} \int_{\mathbf{R}^{2n}} (A\phi_{z,\zeta})(x)\overline{\phi_{z,\zeta}(y)}dz\,d\zeta.$$

Hint: Just write $A = AT^*T$ and observe that

$$Tu(z,\zeta) = (2\pi h)^{-n/2} \int \overline{\phi_{z,\zeta}(y)}u(y)dy.$$

Note: For other applications of coherent states one may consult, e.g., [CoRo], [PaUr], and references therein.

2. **Range of T** - On $L^2(\mathbf{R}^{2n})$ we consider the operator

$$\Pi := TT^*,$$

where T is the FBI transform defined in (3.1.1).

(i) Prove that $\Pi^2 = \Pi = \Pi^*$ and $T^*\Pi = T^*$, and deduce that Π is the orthogonal projector onto $T(L^2(\mathbf{R}^n))$.

(ii) Let $v \in L^2(\mathbf{R}^{2n})$ be of the form

$$v(x,\xi) = e^{-\xi^2/2h}a(x-i\xi),$$

where a is an entire function on $\mathbf{C}^n$. Prove that

$$\begin{aligned}&\Pi v(x,\xi)\\ &= \frac{1}{(2\pi h)^n}\int e^{-[(y-x)^2+(\eta-\xi)^2]/4h-[\eta^2+i(y-x)(\eta+\xi)]/2h}a(y-i\eta)\,dy\,d\eta.\end{aligned}$$

(iii) Making (and justifying) the change of contour of integration

$$\mathbf{R}^{2n} \ni (y,\eta) \mapsto (y+i\eta-i\xi,\eta),$$

show that $\Pi v = v$. (Hint: Interpret the previous integral as an oscillatory one, e.g., passing through the limit when $\varepsilon \to 0_+$ of the integral obtained by multiplying the integrated function with $e^{-\varepsilon\eta^2}$.)

(iv) Deduce from the previous questions that

$$T(L^2(\mathbf{R}^n)) = L^2(\mathbf{R}^{2n}) \cap e^{-\xi^2/2h}\mathcal{H}(\mathbf{C}^n_{x-i\xi}),$$

where $\mathcal{H}(\mathbf{C}^n_{x-i\xi})$ denotes the space of entire functions of $x-i\xi$ on $\mathbf{C}^n$.

(v) Following the same procedure, prove that

$$T(\mathcal{S}'(\mathbf{R}^n)) = \mathcal{S}'(\mathbf{R}^{2n}) \cap e^{-\xi^2/2h}\mathcal{H}(\mathbf{C}^n_{x-i\xi}).$$

3. **Frequency Set** - Using Definition 2.9.1, prove that a point $(x_0, \xi_0) \in \mathbf{R}^{2n}$ is not in the frequency set of $u \in L^2(\mathbf{R}^n)$ if and only if $Tu(x,\xi) = \mathcal{O}(h^\infty)$ uniformly in a neighborhood of (x_0, ξ_0).

   Hint: First prove that if $\chi = \chi(y,\eta) \in S_{2n}(1)$ vanishes near some point (y_0, η_0), then $\chi(y, hD_y)(e^{i(y-x)\xi/h-(x-y)^2/2h}) = \mathcal{O}(h^\infty)$ for (x,ξ) close enough to (y_0,η_0) (here $\chi(y,hD_y)$ denotes any quantization of χ). Then for the necessary condition write $u(y) = \chi(y,hD_y)u + (1 - \chi(y,hD_y))u$ in the expression of Tu, and for the sufficient condition just use that $\mathrm{Op}_h^W(\chi)u = \mathrm{Op}_h^W(\chi)T^*(Tu)$.

4. **Weighted Estimates in the General C^∞ Case** - Let $p \in S_{2n}(1)$. By using an almost-analytic extension of p (see Exercise 23 of Chapter 2), prove an estimate similar to the one of Corollary 3.5.3 but with $e^{\psi/h}$ replaced by h^ψ.
   Hint: Just mimic the proof, working with $(h\ln h)\psi$ instead of ψ, and use the Stokes formula to justify modulo $\mathcal{O}(h^\infty)$ the change of contour given in (3.5.16) directly with $\nu = -i$. The final estimate is

$$\left\langle f h^\psi TPu, h^\psi Tv\right\rangle = \left\langle (\widetilde{p}(x,\xi;h) + R(h))h^\psi Tu, h^\psi Tv\right\rangle$$

   with $\|R(h)\| = \mathcal{O}(h^\infty)$ and

$$\widetilde{p}(x,\xi;h) \sim f(x,\xi)p_a(x - 2h\ln h\partial_z\psi, \xi + 2ih\ln h\partial_z\psi) + \sum_{k\geq 1} h^h\widetilde{p}_k(x,\xi,h)$$

   in $S_{2n}(1)$, where p_a denotes an almost-analytic extension of p.

5. **Weighted Estimates in the Gevrey Case** - Assume now that $p \in S_{2n}(1)$ is s-*Gevrey* for some $s > 1$, that is,

$$\sup_{\mathbf{R}^{2n}} |\partial^\alpha p| \leq C^{1+|\alpha|}(\alpha!)^s$$

   for all $\alpha \in \mathbf{N}^{2n}$ and for some constant $C > 0$. Prove again an estimate similar to the one of Corollary 3.5.3 but with $e^{\psi/h}$ replaced by $e^{\psi/h^{\frac{1}{s}}}$.

6. **Levi–Mizohata Uniqueness Theorem** - In 1 dimension, one considers for $h > 0$ the differential operator $A = hD_t + it$.

(i) Determinate all the solutions $u \in \mathcal{D}'(\mathbf{R})$ of the equation $Au = 0$, and show that the only one that belongs to $L^2(\mathbf{R})$ is the trivial one. Is it still true when A is replaced by $B = hD_t - it$?

(ii) Now we try to find a generalization of this result. Let $p = p(x,\xi) \in S_{2n}(1)$, and denote $p_1 = \mathrm{Re}p$ and $p_2 = \mathrm{Im}p$. We assume that there exists some $\delta > 0$ such that for all $(x,\xi) \in \mathbf{R}^{2n}$ the following implication is true:

$$|p(x,\xi)| \leq \delta \Longrightarrow \{p_1, p_2\}(x,\xi) \geq \delta, \tag{3.6.1}$$

where $\{.,.\}$ is the Poisson bracket. We set $P = Op_h^W(p)$ and $P_j = Op_h^W(p_j)$ $(j = 1,2)$, and we denote by T the global FBI transform defined in (3.1.1).

(ii.a) Prove the a priori estimate

$$\|Pu\|^2 = \|P_1u\|^2 + \|P_2u\|^2 + h\langle\{p_1,p_2\}Tu, Tu\rangle_{L^2(\mathbf{R}^{2n})} + \mathcal{O}(h^2\|u\|^2)$$

uniformly with respect to $h > 0$ small enough and $u \in L^2(\mathbf{R}^n)$. (The norms without index are those in $L^2(\mathbf{R}^n)$.)

(ii.b) We set $\Sigma_\delta = \{(x,\xi) \in \mathbf{R}^{2n}\ ;\ |p(x,\xi)| \leq \delta\}$, and Σ_δ^C its complement in $\mathbf{R}^{2n}$. Show that if p satisfies (3.6.1), then there exists $C_\delta > 0$ such that for all $h > 0$ and for all $u \in L^2(\mathbf{R}^n)$ one has

$$\|P_1u\|^2 + \|P_2u\|^2 + h\langle\{p_1,p_2\}Tu, Tu\rangle_{L^2(\mathbf{R}^{2n})}$$

$$\geq \max\left\{\frac{1}{C_\delta}\|Tu\|^2_{\Sigma_\delta^C} - C_\delta h\|Tu\|^2_{\Sigma_\delta}\ ;\ \frac{h}{C_\delta}\|Tu\|^2_{\Sigma_\delta} - C_\delta h\|Tu\|^2_{\Sigma_\delta^C}\right\}.$$

(ii.c) Deduce from (ii.a) and (ii.b) that if p satisfies (3.6.1), then for $h > 0$ small enough one has the implication

$$\left.\begin{array}{r} u \in L^2(\mathbf{R}^n) \\ Pu = 0 \end{array}\right\} \Longrightarrow u = 0. \tag{3.6.2}$$

(ii.d) Let $Q = hD_t + itOp^W(a)$ with $a = a(x,\xi) \in S_{2n}(1)$ elliptic and positive. Reducing to $S_{2n}(1)$ by multiplying Q with a convenient operator, show that the implication (3.6.2) with P replaced by Q (and n replaced by $n+1$) is true.

7. Calculate the second term $\widetilde{p}_1$ in Corollary 3.5.7 when $t = \frac{1}{2}$. (Hint: Notice that when $t = \frac{1}{2}$, then $r_2 = 0$ in the proof of Theorem 3.5.1. The final result is: $\widetilde{p}_1 = -\frac{1}{4}\Delta p$.)

8. **Agmon Estimates** - Let $p \in S_{2n}(1)$ be such that $p(x,\xi)$ extends holomorphycally with respect to ξ near $\{\xi \in \mathbf{C}^n \ ; \ |\mathrm{Im}\xi| \leq c_0\}$ for some $c_0 > 0$, and remains bounded there together with all its derivatives. Let also $\varphi = \varphi(x)$ be a real-valued smooth function on $\mathbf{R}^n$, bounded together with all its derivatives, and satisfying $|\nabla\varphi| \leq c_0$. We set $P = \mathrm{Op}_h^W(p)$.

   (i) Making a change of contour of integration, prove that the operator $P_\varphi := e^{\varphi/h}Pe^{-\varphi/h}$ is an h-pseudodifferential operator with symbol $p_\varphi(x,\xi,h) = p(x,\xi + i\nabla\varphi(x)) + \mathcal{O}(h)$ in $S_{2n}(1)$.

   (ii) Using Corollary 3.5.7, deduce from (i) that for all $u \in L^2(\mathbf{R}^n)$ one has

   $$\left\langle e^{\varphi/h}Pu, e^{\varphi/h}u\right\rangle = \left\langle p(x,\xi + i\nabla\varphi(x))Te^{\varphi/h}u, Te^{\varphi/h}u\right\rangle + \mathcal{O}(h\|e^{\varphi/h}u\|^2)$$

   uniformly with respect to h small enough and $u \in L^2(\mathbf{R}^n)$. (Hint: Just rewrite $\left\langle e^{\varphi/h}Pu, e^{\varphi/h}u\right\rangle = \langle P_\varphi v, v\rangle$ with $v = e^{\varphi/h}u$.)

   (iii) Give a generalization of the previous estimate when $p \in S_{2n}(\langle\xi\rangle^m)$ with $m \geq 1$.

   (iv) In the particular case of the Schrödinger operator $P_V = -h^2\Delta + V(x)$ with $V \in S_n(1)$, deduce for h small enough the following inequality:

   $$\mathrm{Re}\left\langle e^{\varphi/h}P_V u, e^{\varphi/h}u\right\rangle \geq \left\langle (V(x) - |\nabla\varphi(x)|^2)e^{\varphi/h}u, e^{\varphi/h}u\right\rangle - Ch\|e^{\varphi/h}u\|^2,$$

   where C is some positive constant.

   (v) Make $e^{\varphi/h}P_V e^{-\varphi/h}$ explicit by a direct computation, and deduce that one actually has the so-called *Agmon inequality*:

   $$\mathrm{Re}\left\langle e^{\varphi/h}P_V u, e^{\varphi/h}u\right\rangle \geq \left\langle \left(V(x) - |\nabla\varphi(x)|^2\right)e^{\varphi/h}u, e^{\varphi/h}u\right\rangle.$$

   (vi) If $u \in L^2(\mathbf{R}^n)$ satisfies $P_V u = Eu$ for some $E \in \mathbf{R}$, and $\|u\| = 1$, then deduce from (v) (or even from (iv)) that for any $\varepsilon > 0$ and any φ such that $|\nabla\varphi(x)|^2 \leq V(x) - E - \varepsilon$ on $\mathrm{Supp}\varphi$, one has $\|e^{\varphi/h}u\| = \mathcal{O}(1)$ uniformly as $h \to 0_+$ (*Agmon estimates*).

Note: This type of estimate has been used by many authors to obtain precise exponential decay of eigenfunctions of Schrödinger operators, see, e.g., [Ag, BrCoDu, Hel, HeSj1, HeSj2, HiSi, Si]. In particular, when V admits a nondegenerate minimum at some point x_0, such estimates make it possible to get the WKB asymptotics near x_0 (i.e., asymptotics of the form $(\sum a_j(x)h^j)e^{-\varphi/h}$ as in Exercise 6 of Chapter 2) of the first eigenfunctions of P_V ([Hel, HeSj1]).

9. Rewrite Theorem 3.5.1 when T is replaced by T_A defined in (3.4.1). In particular, calculate $\widetilde{p}_0$ of Corollary 3.5.3 explicitly when $A = \mu I$ with $\mu > 0$.

   Answer: $\widetilde{q}_0(x,\xi)$ becomes

   $$q(x,\xi,\xi-(A_1+A_2A_1^{-1}A_2)\partial_\xi\psi - A_2A_1^{-1}\partial_x\psi, A_1^{-1}\partial_x\psi + A_1^{-1}A_2\partial_\xi\psi),$$

   and when $A = \mu I$, then $\widetilde{p}_0$ becomes

   $$p(x-\mu^{-1}\partial_x\psi - i\partial_\xi\psi, \xi + i\partial_x\psi - \mu\partial_\xi\psi).$$

   Note: In the case where $\psi = \psi(x)$ does not depend on ξ, and μ is taken very large, then the previous $\widetilde{p}_0$ becomes arbitrarily close to $p(x,\xi+i\partial_x\psi)$, which is the quantity appearing in the Agmon estimates (see Exercise 8 above).

10. **Semiclassical Measures** - Let $u = (u_h)_{h\in(0,h_0]}$ be a family of functions in $L^2(\mathbf{R}^n)$ such that $\|u_h\|_{L^2} = 1$ (so that $|T_hu_h(x,\xi)|^2dxd\xi$ is a probability measure on $L^2(\mathbf{R}^{2n})$, where T_h denotes the h-dependent FBI transform defined in (3.1.1)). An (h-independent) probability measure $d\mu$ on $\mathbf{R}^{2n}$ is called a *semiclassical measure of* u if there exists some sequence $(h_j)_{j\in\mathbf{N}}$ converging to 0_+ such that

    $$|T_{h_j}u_{h_j}(x,\xi)|^2dxd\xi \to d\mu \text{ weakly, as } j\to+\infty.$$

    (The function $|Tu(x,\xi)|^2$ is called the *Husimi function* attached to u.)

    (i) Let $d\mu$ be a semiclassical measure of u, corresponding to a sequence $(h_j)_{j\in\mathbf{N}}$ converging to 0_+, and let $A(x,hD_x)$ denote a semiclassical pseudodifferential operator admitting a principal symbol $a_0\in S_{2n}(1)$. Using Corollary 3.5.3 with $\psi = 0$, prove that

    $$\left\langle A(x,h_jD_x)u_{h_j},u_{h_j}\right\rangle_{L^2} \to \int_{\mathbf{R}^{2n}} a_0(x,\xi)d\mu(x,\xi) \qquad (j\to+\infty). \tag{3.6.3}$$

Conversely, prove that if the property (3.6.3) is true for all $a \in S_{2n}(1)$, then $d\mu$ is a semiclassical measure of u.

(ii) Assume that u is a solution of the equation $Pu = 0$, where $P = P(x, hD_x)$ is a semiclassical pseudodifferential operator admitting a principal symbol $p_0 \in S_{2n}(1)$. Applying (i) with $A(x, hD_x) \circ P(x, hD_x)$ instead of $A(x, hD_x)$, prove that if $d\mu$ is a semiclassical measure of u, then it satisfies $p_0 d\mu = 0$.

(iii) Assume now that $u_h = u_h(t) \in C^1(\mathbf{R}_t\ ;\ L^2(\mathbf{R}^n))$ is solution of

$$\begin{cases} (hD_t + P_h)u_h = 0, \\ u_h\,|_{t=0} = v_h, \end{cases}$$

with $P_h = \mathrm{Op}_h^W(p)$, $p \in S_{2n}(1)$ real-valued, $p = p_0 + \mathcal{O}(h)$ in $S_{2n}(1)$ (p_0 independent of h), and $\|v_h\|_{L^2} = 1$. Assume, moreover, that v admits a semiclassical measure $d\nu$ corresponding to a sequence $(h_j)_{j\in\mathbf{N}}$. Then prove that for all $t \in \mathbf{R}$, $u(t)$ admits a semiclassical measure $d\mu_t$ corresponding to the same sequence $(h_j)_{j\in\mathbf{N}}$, and that it satisfies

$$\begin{cases} \partial_t(d\mu_t) + \{p_0, d\mu_t\} = 0, \\ d\mu_0 = d\nu. \end{cases} \tag{3.6.4}$$

(Hint: Use the equation to show that for all pseudodifferential operator $A = A(x, hD_x)$ one has $hD_t\,\langle Au, u\rangle = \langle [P_h, A]u, u\rangle$, and deduce that any weak limit $d\mu_t$ of a subsequence of $|T_{h_j}u_{h_j}(x,\xi)|^2 dx\,d\xi$ is a solution of (3.6.4); then conclude by observing that the system (3.6.4) admits a unique solution.)

(iv) In the particular case where u_h (independent of t) is a solution of $P_h u_h = 0$, deduce from (iii) a result of propagation for the semiclassical measures of u.

Note: For further results on the semiclassical measures, one may consult, e.g., [GeP].

11. **Lagrangian States** - An h-dependent function $u \in L^2(\mathbf{R}^n)$ is called *Lagrangian* if it can be written in the form $I_\varphi(a)$ defined in (2.4.8) with the conditions (2.4.6) and (2.4.7).

(i) In this case, prove that

$$\mathrm{FS}(u) \subset \{(x, \nabla_x\varphi(x,\theta))\ ;\ \nabla_\theta\varphi(x,\theta) = 0\}.$$

(ii) If, moreover, $a \in S^{hol}_{n+n'}(\langle\theta\rangle^m)$ and $\varphi(x,\theta)$ is analytic in a complex strip around $\mathbf{R}^{n+n'}$, prove that

$$\mathrm{MS}(u) \subset \{(x, \nabla_x\varphi(x,\theta))\ ;\ \nabla_\theta\varphi(x,\theta) = 0\}.$$

Hint: In the expression of $T_\mu u(x,\xi)$ with $\mu > 0$ large enough, write that $\varphi(y,\theta) = \varphi(x,\theta) + (y-x)\nabla_x\varphi(x,\theta) + \mathcal{O}(|x-y|^2)$; then use the operator $L = \left(1 + \frac{|\nabla_\theta\varphi|^2}{h^2}\right)^{-1}\left(1 + \frac{1}{h}\nabla_\theta\varphi.D_\theta\right)$ to make integrations by parts near a point x for which $\nabla_\theta\varphi(x,\theta) \neq 0$ for all θ; finally, make integrations by parts with respect to y in the C^∞ case, or make a change of contour of integration with respect to y in the analytic case, when $\xi \neq \nabla_x\varphi(x,\theta)$.

Note: $I_\varphi(a) \in L^2$ if, e.g., $|\nabla_\theta\varphi| \geq (\langle\theta\rangle^\rho + \langle x\rangle^\rho)/C$, or if $a = a(x) \in L^2$.

12. Give a result of microlocal decay at infinity, by using a sequence of bounded weight functions $(\psi_j)_{j\in\mathbf{N}}$ converging at infinity toward an unbounded function ψ (but with bounded gradient).

13. Do the same as in the previous exercise, but this time assuming only that $\mathrm{Hess}\,\psi$ is uniformly bounded (together with all its derivatives) and sufficiently small, while p is holomorphic (and admits symbol-type estimates) in a complex sector of the form

$$\mathcal{S}_\delta = \{(x,\xi) \in \mathbf{C}^{2n}\ ;\ |\mathrm{Im}\,(x,\xi)| < \delta\,\langle \mathrm{Re}(x,\xi)\rangle\}$$

with $\delta > 0$.

Chapter 4

Applications to the Solutions of Analytic Linear PDEs

The purpose of this chapter is to apply the results of the previous sections for finding out many microlocal properties of the L^2-solutions of partial differential equations of the type

$$P(x, hD_x)u = 0,$$

where $P = \sum_{|\alpha|\leq m} a_\alpha(x)(hD_x)^\alpha$ has analytic coefficients. In particular, properties of localization and propagation are given for the microsupport $\mathrm{MS}(u)$ when u is normalized by $\|u\|_{L^2} = 1$.

4.1 Projection of the Microsupport

Before studying the microlocal properties of the solutions of analytic PDEs, in this section we explain how one can recover some *local* properties of $u \in L^2(\mathbf{R}^n)$ from knowledge of $\mathrm{MS}(u)$.

Definition 4.1.1 *For $u \in L^2(\mathbf{R}^n)$ (depending also on h), we say that u is* **exponentially small** *near some $x_0 \in \mathbf{R}^n$ if there exist a neighborhood U of x_0 in $\mathbf{R}^n$ and $\delta > 0$ such that*

$$\|u\|_{L^2(U)} = \mathcal{O}\left(e^{-\delta/h}\right). \tag{4.1.1}$$

We denote by $\mathrm{Suppexp}\, u$ *(the* **exponential support** *of u) the complement in $\mathbf{R}^n$ of those points near which u is exponentially small.*

In particular, Suppexp$u$ is a closed subset of $\mathbf{R}^n$.

Still denoting by T the FBI transform introduced in the previous chapter, we then have the following theorem:

Theorem 4.1.2 *Let $u \in L^2(\mathbf{R}^n)$ such that $\|u\|_{L^2} \leq 1$, and $x_0 \in \mathbf{R}^n$. Then $x_0 \notin$ Suppexp$u$ if and only if there exist a neighborhood V of x_0 and $\delta_1 > 0$ such that*

$$\|Tu\|_{L^2(V\times\mathbf{R}^n)} = \mathcal{O}(e^{-\delta_1/h}). \tag{4.1.2}$$

Proof Assume first that $x_0 \notin$ Suppexp$u$. Then there exist a bounded neighborhood U of x_0 and $\delta > 0$ such that (4.1.1) is satisfied, and if V is another neighborhood of x_0 such that $\overline{V} \subset \overset{\circ}{U}$, one has

$$\begin{aligned}
&\|Tu\|^2_{L^2(V\times\mathbf{R}^n)}\\
&= |\alpha_{n,h}|^2 \int_{x\in V} e^{i(y'-y)\xi/h-(x-y)^2/2h-(x-y')^2/2h}u(y)\overline{u(y')}dy\,dy'\,dx\,d\xi\\
&= (\pi h)^{-n/2}\int_{x\in V} e^{-(x-y)^2/h}|u(y)|^2dy\,dx\\
&= (\pi h)^{-n/2}\int_{\substack{x\in V\\ y\in U}} e^{-(x-y)^2/h}|u(y)|^2dy\,dx\\
&\quad +(\pi h)^{-n/2}\int_{\substack{x\in V\\ y\notin U}} e^{-(x-y)^2/h}|u(y)|^2dy\,dx,
\end{aligned}$$

and thus, by (4.1.1) and the Cauchy–Schwarz inequality,

$$\|Tu\|^2_{L^2(V\times\mathbf{R}^n)} = \mathcal{O}(e^{-2\delta/h} + h^{-n/2}e^{-\varepsilon/h})$$

with $\varepsilon = \inf\limits_{\substack{x\in V\\ y\notin U}} |x-y|^2$, which by construction is a positive number. In particular, for $h > 0$ small enough we get (4.1.2) with $\delta_1 = \min\{2\delta, \frac{\varepsilon}{2}\}$.

Conversely, assume that (4.1.2) is satisfied, and denote by U a bounded open neighborhood of x_0 such that $\overline{U} \subset \overset{\circ}{V}$. Let also $\chi \in C_0^\infty(U)$ such that $\chi = 1$ near x_0. Writing

$$u(y) = T^*Tu(y) = \alpha_{n,h}\int e^{-i(x-y)\xi/h-(x-y)^2/2h}Tu(x,\xi)dx\,d\xi,$$

we have

$$\begin{aligned}
\|\chi u\|^2 &= |\alpha_{n.h}|^2\int_{y\in W'} e^{iy(\xi-\xi')/h+i(x'\xi'-x\xi)/h-(x-y)^2/2h-(x'-y)^2/2h}|\chi(y)|^2\\
&\qquad \times Tu(x,\xi)\overline{Tu(x',\xi')}dx\,d\xi\,dx'\,d\xi'\,dy.
\end{aligned}$$

Set

$$L = \frac{1}{1+|\xi-\xi'|^2}(1+h(\xi-\xi')\cdot D_y).$$

Then $L(e^{iy(\xi-\xi')/h}) = e^{iy(\xi-\xi')/h}$, and therefore, using L to make N integrations by parts (for some N large enough), we get

$$\begin{aligned}\|\chi u\|^2 \\ = \ |\alpha_{n.h}|^2 \int_{y\in W'} e^{iy(\xi-\xi')/h+i(x'\xi'-x\xi)/h}({}^tL)^N \left(e^{-(x-y)^2/2h-(x'-y)^2/2h}|\chi(y)|^2\right) \\ \times Tu(x,\xi)\overline{Tu(x',\xi')}dx\,d\xi\,dx'\,d\xi'\,dy. \qquad (4.1.3)\end{aligned}$$

Now we notice that

$$({}^tL)^N\left(e^{-(x-y)^2/2h-(x'-y)^2/2h}|\chi(y)|^2\right) = \mathcal{O}\left(\frac{1}{\langle\xi-\xi'\rangle^N}e^{-(x-y)^2/4h-(x'-y)^2/4h}\right),$$

and it is supported in $\{y\in U\}$. As a consequence, we deduce from (4.1.3) that

$$\begin{aligned}\|\chi u\|^2 \\ = \mathcal{O}\left(h^{-3n/2}\int_{y\in U}\frac{e^{-(x-y)^2/4h-(x'-y)^2/4h}}{\langle\xi-\xi'\rangle^N}|Tu(x,\xi)|\cdot|Tu(x',\xi')|dx\,d\xi\,dx'\,d\xi'\,dy\right). \\ (4.1.4)\end{aligned}$$

Using Schur's lemma (see Section 2.8), we see that if $N>n$, then $\langle\xi-\xi'\rangle^{-N}$ is the distribution kernel of a bounded operator on $L^2(\mathbf{R}^n_\xi)$. In particular, for any $f,g\in L^2(\mathbf{R}^n_\xi)$ one has

$$\int\langle\xi-\xi'\rangle^{-N}|f(\xi)g(\xi')|d\xi d\xi' = \mathcal{O}(\|f\|_{L^2}\|g\|_{L^2}),$$

and thus we deduce from (4.1.4) that

$$\begin{aligned}\|\chi u\|^2 \ = \ \mathcal{O}\left(h^{-3n/2}\right)\int_{y\in U} e^{-(x-y)^2/4h-(x'-y)^2/4h} \qquad (4.1.5) \\ \times\left(\int|Tu(x,\xi)|^2d\xi\int|Tu(x',\xi)|^2d\xi\right)^{1/2}dx\,dx'\,dy.\end{aligned}$$

Finally, cutting the domain of integration in $\{x,x'\in V\}\cup\{x \text{ or } x'\notin V\}$, and using (4.1.2), we get from (4.1.5),

$$\|\chi u\|^2 = \mathcal{O}(e^{-2\delta_1/h}+h^{-3n/2}e^{-\delta_2/h})$$

with $\delta_2 = \inf_{\substack{y\in U\\ x\notin V}} |x-y|^2/4 > 0$. In particular, $\|u\|_{L^2(\mathrm{Supp}\chi)}$ is exponentially small, and therefore $x_0 \notin \mathrm{Supp}\exp u$. ◇

We deduce immediately from this theorem the following corollary:

Corollary 4.1.3 *Let $u \in L^2(\mathbf{R}^n)$ such that $\|u\|_{L^2} \leq 1$, and $x_0 \in \mathbf{R}^n$.*
(i) If $x_0 \notin \mathrm{Supp}\exp u$, then $\{x_0\} \times \mathbf{R}^n \cap \mathrm{MS}(u) = \emptyset$.
(ii) Conversely, if $\{x_0\} \times \mathbf{R}^n \cap \mathrm{MS}(u) = \emptyset$ and if there exist $C_0, \delta > 0$ and a neighborhood U of x_0 such that

$$\|Tu\|_{L^2(U\times\{|\xi|\geq C_0\})} = \mathcal{O}(e^{-\delta/h}), \tag{4.1.6}$$

then $x_0 \notin \mathrm{Supp}\exp u$.
(iii) In particular, if the property (4.1.6) is satisfied for any $x_0 \in \mathbf{R}^n$ such that $\{x_0\} \times \mathbf{R}^n \cap \mathrm{MS}(u) = \emptyset$, then

$$\boxed{\Pi_x(\mathrm{MS}(u)) = \mathrm{Supp}\exp u,} \tag{4.1.7}$$

where Π_x denotes the projection $\mathbf{R}^{2n} \ni (x,\xi) \mapsto x$.

As one can see, the identity (4.1.7) requires a little bit more than the local exponential decay of Tu in the definition of $\mathbf{R}^{2n}\backslash\mathrm{MS}(u)$; namely, one needs some uniformity in this decay when $|\xi|$ becomes large. This uniformity is, of course, not always satisfied, as can be seen in the following example:

Example - For $x \in \mathbf{R}$, let $u(x;h) = (\pi h)^{-1/2}e^{ix/h^2 - x^2/2h}$. Then $\|u\|_{L^2} = 1$ and $0 \in \mathrm{Supp}\exp u$, but $\{0\} \times \mathbf{R} \cap \mathrm{MS}(u) = \emptyset$ (see Exercise 1 of this chapter).

The fundamental reason why the previous example does not satisfy (4.1.6) is that u oscillates too much near 0, and therefore its Fourier transform is concentrated near infinity. However, as we shall see now, this is not the case for the solutions of some types of analytic semiclassical PDEs (such as the Schrödinger operator), which have the effect of localizing with respect to the momentum variable ξ. These equations will be called *classically elliptic* in the following sense:

Definition 4.1.4 *Let $m \in \mathbf{R}$. We denote by $S_{2n}^{hol}(\langle\xi\rangle^m)$ the space of those $p \in S_{2n}(\langle\xi\rangle^m)$ that can be extended holomorphically in a complex band $\Sigma_a =: \{(x,\xi) \in \mathbf{C}^{2n}\ ;\ |\mathrm{Im}x| + |\mathrm{Im}\xi| < a\}$ for some positive a (independent of h).*

For $p \in S_{2n}^{hol}(\langle\xi\rangle^m)$ and $x_0 \in \mathbf{R}^n$, we say that p is **classically elliptic near** *x_0 if there exists $C_1 > 0$ such that*

$$|p(x,\xi)| \geq \frac{1}{C_1} \langle \mathrm{Re}\, \xi\rangle^m \tag{4.1.8}$$

uniformly with respect to $\mathrm{Re}\, x$ in a neighborhood of x_0, $|\mathrm{Re}\, \xi| \geq C_1$, and $|\mathrm{Im}\, x|$, $|\mathrm{Im}\, \xi|$ and h small enough.

In this definition (as well as in the sequel of this chapter), we did not write down explicitly the dependence of p with respect to h, although this dependence is always possible in applications.

Then we have the following result:

Theorem 4.1.5 *Let $p \in S_{2n}^{hol}(\langle\xi\rangle^m)$ be classically elliptic in the sense of the previous definition, and let $u \in L^2(\mathbf{R}^n)$ satisfy, for some $t \in [0,1]$,*

$$\begin{cases} \mathrm{Op}_h^t(p)u = 0, \\ \|u\|_{L^2} \leq 1. \end{cases} \tag{4.1.9}$$

Then one has

$$\Pi_x(\mathrm{MS}(u)) = \mathrm{Supp}\exp u.$$

Remark 4.1.6 An example is given by the Schrödinger operator: $P = -h^2\Delta + V(x)$ with $V \in S_n^{hol}(1)$ (in this case, one has $m = 2$).

Proof In view of Corollary 4.1.3, it is sufficient to show that for any $x_0 \in \mathbf{R}^n$, if $\{x_0\} \times \mathbf{R}^n \cap \mathrm{MS}(u) = \emptyset$, then there exist a neighborhood U of x_0 and $C_0 > 0$ such that (4.1.6) is valid. Given such a point x_0, let $a > 0$ be small enough that p extends holomorphically in Σ_a, and let C_1 be the positive constant of Definition 4.1.4. Then since $\{x_0\} \times \{|\xi| \leq C_1\}$ is a compact set that does not meet $\mathrm{MS}(u)$, there exist a neighborhood V of x_0 and $\delta > 0$ such that

$$\|Tu\|_{L^2(V\times\{|\xi|\leq C_1\})} = \mathcal{O}(e^{-\delta/h}). \tag{4.1.10}$$

Then let $\psi = \psi(x) \in C^\infty(\mathbf{R}^n)$ be a real-valued function satisfying

$$\begin{cases} \mathrm{Supp}\psi \subset V; \\ \sup\limits_{\mathbf{R}^n} |\partial_x \psi| < a; \\ \psi \leq \delta \text{ everywhere}; \\ \psi(x_0) > 0. \end{cases}$$

Applying the microlocal exponential estimate given in Corollary 3.5.5 with $P = \langle hD_x\rangle^{-m}\operatorname{Op}_h^t(p)$, we get

$$\|e^{\psi(x)/h}TPu\|^2 = \left\|\frac{p(x-\partial_x\psi,\xi+i\partial_x\psi)}{\langle\xi+i\partial_x\psi\rangle^m}e^{\psi/h}Tu\right\|^2 + \mathcal{O}(h)\|e^{\psi/h}Tu\|^2,$$

and therefore, since by assumption $Pu = 0$,

$$\left\|\frac{p(x-\partial_x\psi,\xi+i\partial_x\psi)}{\langle\xi+i\partial_x\psi\rangle^m}e^{\psi/h}Tu\right\|^2 = \mathcal{O}(h)\|e^{\psi/h}Tu\|^2. \tag{4.1.11}$$

Since, moreover, (by (4.1.8))

$$\left|\frac{p(x-\partial_x\psi,\xi+i\partial_x\psi)}{\langle\xi+i\partial_x\psi\rangle^m}\right| \geq \frac{1}{2C_1}$$

for $|\xi| \geq C_1$, we deduce from (4.1.11) that

$$\|e^{\psi/h}Tu\|^2_{L^2(\mathbf{R}^n\times\{|\xi|\geq C_1\})} = \mathcal{O}(h)\|e^{\psi/h}Tu\|^2,$$

and therefore, for h small enough,

$$\left\|e^{\psi/h}Tu\right\|^2_{L^2(\mathbf{R}^n\times\{|\xi|\geq C_1\})} = \mathcal{O}(h)\left\|e^{\psi/h}Tu\right\|^2_{L^2(\mathbf{R}^n\times\{|\xi|\leq C_1\})}.$$

But then, using (4.1.10) and the properties of ψ, this gives, in particular,

$$\|e^{\psi/h}Tu\|^2_{L^2(\mathbf{R}^n\times\{|\xi|\geq C_1\})} = \mathcal{O}(1),$$

and thus, setting $U = \{x\in V\ ;\ \psi(x)\geq\psi(x_0)/2\}$ (which is a neighborhood of x_0),

$$\|Tu\|^2_{L^2(U\times\{|\xi|\geq C_1\})} = \mathcal{O}(e^{-\psi(x_0)/2h}),$$

from which the result follows by virtue of Corollary 4.1.3. ◇

Remark 4.1.7 In this proof we have taken $\psi = \psi(x)$ independent of ξ. As a consequence, we did not use the analyticity of p with respect to x, and in fact, the result remains valid for those p that can be extended holomorphically with respect to the ξ-variables only. In particular, the result is actually true for the Schrödinger operator $P = -h^2\Delta + V(x)$ with $V \in S_n(1)$.

Remark 4.1.8 As one can see from the proof, the result remains valid if we replace the first equation of (4.1.9) by

$$\mathrm{Op}_h^t(p)u = \varepsilon(h)\mathrm{Op}_h^t(r)u$$

with $r \in S_{2n}^{hol}(\langle\xi\rangle^m)$ and $\varepsilon(h) \to 0$ as $h \to 0$. This reflects the fact that only the "principal" symbol of $\mathrm{Op}_h^t(p)$ (that is, the symbol modulo errors tending to zero with h) is actually involved in the definition of ellipticity. Moreover, and as usual in the semiclassical theory, the order function $\langle\xi\rangle^m$ can be replaced by any other order function on $\mathbf{R}^{2n}$.

Remark 4.1.9 One can see in the same way that if $p \in S_{2n}^{hol}(1)$ satisfies $|p(x,\xi)| \geq \dfrac{1}{C_0}$ for all $|x| \geq C_0$ and $\xi \in \mathbf{R}^n$, then the solutions u of (4.1.9) satisfy

$$\boxed{\Pi_\xi(\mathrm{MS}(u)) = \mathrm{Supp\,exp}(\mathcal{F}_h u),}$$

where $\Pi_\xi \ : \ \mathbf{R}^{2n} \ni (x,\xi) \mapsto \xi$.

In fact, this is a consequence of the identity (see Proposition 3.4.3)

$$Tu(x,\xi) = e^{ix\xi/h}(T\mathcal{F}_h u)(\xi, -x)$$

and of the elementary fact that

$$\mathcal{F}_h \mathrm{Op}_h^0(p)\mathcal{F}_h^{-1}v(\xi) = \mathrm{Op}_h^1(p(-\xi^*,\xi))v(\xi).$$

Then if $\mathrm{Op}_h^t(p)u = 0$, we have $\mathrm{Op}_h^1(q)\mathcal{F}_h u(\xi) = 0$ with $q(\xi,\xi^*) = p(-\xi^*,\xi) + \mathcal{O}(h)$ in $S_{2n}^{hol}(1)$ (see, e.g., Exercise 4 of Chapter 2), and the result follows by applying Theorem 4.1.5 with q and $\mathcal{F}_h u$ instead of p and u.

4.2 Characteristic Set and Microsupport

In this section we give a first result of localization for the microsupport of solutions of analytic PDEs, related to the so-called *characteristic set* of an operator, and defined as follows:

Definition 4.2.1 *Let* $p = p(x,\xi,h) \in S_{2n}(\langle\xi\rangle^m)$ *of the form*

$$p(x,\xi;h) = p_0(x,\xi) + \varepsilon(h)r(x,\xi;h)$$

with $p_0 \in S_{2n}(\langle\xi\rangle^m)$ *independent of* h, $r \in S_{2n}(\langle\xi\rangle^m)$ *and* $\varepsilon(h) \in \mathbf{C}$ *tending to* 0 *as* $h \to 0$. *For* $t \in [0,1]$ *given arbitrarily, set* $P = \mathrm{Op}_h^t(p)$. *Then the set*

$$\mathrm{Char}\,(P) =: \{(x,\xi) \in \mathbf{R}^{2n}\ ;\ p_0(x,\xi) = 0\}$$

is called the **characteristic set** *of* P *(or sometimes also of* p*).*

Notice that such a definition is particularly adapted for *classical* symbols, in the sense of Section 2.7.

Then we have the following result:

Theorem 4.2.2 *Let* $p = p(x,\xi,h) \in S_{2n}^{hol}(\langle\xi\rangle^m)$ *be of the same form as in the previous definition and let* $u \in L^2(\mathbf{R}^n)$ *satisfy*

$$\begin{cases} Pu = 0 \\ \|u\|_{L^2} \leq 1 \end{cases}$$

with $P = \mathrm{Op}_h^t(p)$ *(*$t \in [0,1]$ *arbitrary). Then one has*

$$\boxed{\mathrm{MS}(u) \subset \mathrm{Char}\,(P).}$$

Remark 4.2.3 Actually, the result remains true if we replace $L^2(\mathbf{R}^n)$ by any Sobolev space $H^s(\mathbf{R}^n)$ ($s \in \mathbf{R}$). In fact, one will then have

$$\mathrm{Op}(\langle\xi\rangle^s)u \in L^2$$

and

$$\mathrm{Op}(\langle\xi\rangle^s)P\mathrm{Op}(\langle\xi\rangle^{-s}) = \mathrm{Op}_h^t(\tilde{p})$$

with $\tilde{p}$ of the same form as p.

Remark 4.2.4 As we shall see in the proof, the condition $Pu = 0$ is even too strong, and it is sufficient indeed that u satisfy $\|Pu\|_{L^2} = \mathcal{O}(e^{-\delta/h})$ for some $\delta > 0$. Also, the condition $\|u\| \leq 1$ may be replaced by $\|u\| \leq h^{-N}$ for some arbitrary $N \in \mathbf{R}$.

Proof of the Theorem First of all, we can reduce to the case $m = 0$ by composing P to the left with $\mathrm{Op}(\langle\xi\rangle^{-m})$.

Fix $(x_0,\xi_0) \notin \mathrm{Char}\,(P)$. Then one must show that $(x_0,\xi_0) \notin \mathrm{MS}(u)$; that is, Tu is exponentially small near (x_0,ξ_0).

Since $p_0(x_0, \xi_0) \neq 0$ and p_0 is holomorphic near (x_0, ξ_0), there exist a complex neighborhood U of (x_0, ξ_0) and a constant $C_0 > 0$ such that $|p_0(x, \xi)| \geq \frac{1}{C_0}$ for $(x, \xi) \in U$. As a consequence, if h is small enough, we also have

$$|p(x, \xi; h)| \geq \frac{1}{2C_0} \text{ for } (x, \xi) \in U. \tag{4.2.1}$$

Then choose $\psi \in S_{2n}(1)$ real-valued such that

$$\begin{cases} \psi = 0 \text{ on } \mathbf{R}^{2n} \backslash U; \\ |\nabla_{x,\xi}\psi| \text{ sufficiently small so that } (x - 2\partial_z\psi, \xi + 2i\partial_z\psi) \in U \text{ if } \\ (x, \xi) \in U \cap \mathbf{R}^{2n}; \\ \psi(x, \xi) = \delta > 0 \text{ in a neighborhood of } (x_0, \xi_0) \end{cases}$$

(where, as in Chapter 3, we have used the notation $\partial_z = \frac{1}{2}(\partial_x + i\partial_\xi)$).

In particular, we get from (4.2.1)

$$|p(x - 2\partial_z\psi, \xi + 2i\partial_z\psi)| \geq \frac{1}{2C_0} \text{ for } (x, \xi) \in U \cap \mathbf{R}^{2n}. \tag{4.2.2}$$

Applying the microlocal exponential estimate given in Corollary 3.5.5, we get here

$$0 = \|e^{\psi/h}TPu\|^2 = \|p(x - 2\partial_z\psi, \xi + 2i\partial_z\psi)e^{\psi/h}Tu\|^2 + \mathcal{O}(h)\,\|e^{\psi/h}Tu\|^2,$$

and thus in particular,

$$\|p(x - 2\partial_z\psi, \xi + 2i\partial_z\psi)\, e^{\psi/h}Tu\|^2_{L^2(U\cap\mathbf{R}^{2n})} = \mathcal{O}(h)\,\|e^{\psi/h}Tu\|^2. \tag{4.2.3}$$

Then, using (4.2.2) and the fact that $\psi = 0$ on $\mathbf{R}^{2n}\backslash U$, we deduce from (4.2.3) that

$$\|e^{\psi/h}Tu\|^2_{L^2(U\cap\mathbf{R}^{2n})} = \mathcal{O}(h)\left(\|Tu\|^2 + \|e^{\psi/h}Tu\|^2_{L^2(U\cap\mathbf{R}^{2n})}\right)$$

and therefore, for h small enough,

$$\|e^{\psi/h}Tu\|^2_{L^2(U\cap\mathbf{R}^{2n})} = \mathcal{O}(h)\|Tu\|^2 = \mathcal{O}(h).$$

In particular, setting $V = \{(x, \xi) \in \mathbf{R}^{2n}\ ;\ \psi(x, \xi) = \delta\} (\subset U)$,

$$\|Tu\|_{L^2(V)} = \mathcal{O}(e^{-\delta/h}). \tag{4.2.4}$$

Since by construction V is a neighborhood of (x_0, ξ_0), the result follows from (4.2.4). ◇

Remark 4.2.5 The method used in this proof is somehow a microlocal version of the so-called *Agmon estimates method* (see the note in Exercise 8 of Chapter 3). In fact, even in our microlocal context this method can be pushed further in order to give much more precise information about the rate of exponential decay of the solutions (see [Mar1, Mar2, Mar3, MaSo, Na2]).

Application

Consider the semiclassical Schrödinger operator $H = -h^2\Delta + V(x)$, with $V \in S_n^{hol}(1)$ real-valued, and let $u \in L^2(\mathbf{R}^n)$ be a normalized eigenfunction of H, that is, there exists $E = E(h) \in \mathbf{R}$ such that

$$\begin{cases} Hu = Eu \\ \|u\|_{L^2} = 1. \end{cases}$$

Assume also that $E(h)$ admits a limit E_0 as $h \to 0$ (where h may actually take its values in a discrete set $\mathcal{J}$ the closure of which contains 0). Then we can apply Theorem 4.2.2 with $P = H - E$, and we obtain

$$\mathrm{MS}(u) \subset \left\{(x,\xi)\ ;\ \xi^2 + V(x) = E_0\right\}.$$

In quantum mechanics this last set is called the *energy surface* or *energy shell* associated with the *energy* E_0. Moreover, we are in a situation where we can also apply Theorem 4.1.5, and this gives, in particular,

$$\mathrm{Supp}\exp u \subset \{x \in \mathbf{R}^n\ ;\ V(x) \le E_0\}, \tag{4.2.5}$$

which is called the *classically allowed region* associated with the energy E_0. In contrast, the set $\{x \in \mathbf{R}^n\ ;\ V(x) > E_0\}$ is called the *classically forbidden region*, and the inclusion (4.2.5) shows that $|u(x)|^2$ (which can be interpreted as the density of probability of presence of a particle) is exponentially small there.

4.3 Propagation of the Microsupport

The purpose of this section is to establish various properties of invariance for the microsupport of solutions of analytic PDEs. As we shall see, these properties are essentially related to the geometry of the characteristic set of the equation.

We start by giving very basic facts on the so-called Hamilton flow of a given real smooth function on $\mathbf{R}^{2n}$.

(a) Hamilton Flow

For $p = p(x, \xi) \in C^\infty(\mathbf{R}^{2n})$ real-valued, we set

$$\begin{aligned} H_p &=: \frac{\partial p}{\partial \xi}\frac{\partial}{\partial x} - \frac{\partial p}{\partial x}\frac{\partial}{\partial \xi} = \sum_{j=1}^{n}\left(\frac{\partial p}{\partial \xi_j}\frac{\partial}{\partial x_j} - \frac{\partial p}{\partial x_j}\frac{\partial}{\partial \xi_j}\right) \\ &\simeq \left(\frac{\partial p}{\partial \xi}, -\frac{\partial p}{\partial x}\right) = H_p(x, \xi) \end{aligned} \tag{4.3.1}$$

which has to be seen as a vector field on $\mathbf{R}^{2n}$, that is, a smooth function: $\mathbf{R}^{2n} \mapsto \mathbf{R}^{2n}$. (The sign $\simeq$ in (4.3.1) just stands for the identification of H_p with its coordinates in the basis $(\frac{\partial}{\partial x}, \frac{\partial}{\partial \xi})$ of the tangent space $T\mathbf{R}^n$ of $\mathbf{R}^n$.)

Definition 4.3.1 *H_p is called the* **Hamilton field** *of p.*

By the Lipschitz theorem, for all $(x_0, \xi_0) \in \mathbf{R}^{2n}$ one can solve, for $t \in \mathbf{R}$ small enough, the differential system (called the *system of Hamilton–Jacobi*)

$$\begin{cases} \dot{x}(t) = \dfrac{\partial p}{\partial \xi}(x(t), \xi(t)); \\ \dot{\xi}(t) = -\dfrac{\partial p}{\partial x}(x(t), \xi(t)); \\ x(0) = x_0 \ ; \ \xi(0) = \xi_0; \end{cases}$$

where the dots stand for differentiation with respect to t. Equivalently,

$$\begin{cases} \partial_t(x(t), \xi(t)) = H_p(x(t), \xi(t)); \\ (x(0), \xi(0)) = (x_0, \xi_0). \end{cases} \tag{4.3.2}$$

Moreover, the solution of (4.3.2) is unique and is denoted by

$$(x(t), \xi(t)) =: (\exp tH_p)(x_0, \xi_0).$$

Definition 4.3.2 *The application*

$$(t, x, \xi) \mapsto (\exp tH_p)(x, \xi)$$

(which is defined for $(x, \xi) \in \mathbf{R}^{2n}$ near some arbitrary (x_0, ξ_0) and for $t \in \mathbf{R}$ small enough) is called the **Hamilton flow** *associated with p.*

We have in particular

$$\boxed{\frac{\partial}{\partial t}\exp tH_p(x,\xi) = H_p(\exp tH_p(x,\xi)).}$$

Example 1 - Let $p(x) = \xi_n$. Then, setting $x = (x', x_n)$, the system of Hamilton–Jacobi becomes

$$\begin{cases} \dot{x}'(t) = 0, \quad \dot{x}_n(t) = 1, \\ \dot{\xi}(t) = 0, \end{cases}$$

and therefore,

$$\exp tH_{\xi_n}(x_0, \xi_0) = ((x_0', x_0^n + t), \xi_0).$$

Example 2 - For $(x,\xi) \in \mathbf{R}^{2n}$, let $p(x,\xi) = \xi^2 + x^2$ (the so-called *harmonic oscillator*). Here the system of Hamilton–Jacobi is

$$\begin{cases} \dot{x}(t) = 2\xi(t), \\ \dot{\xi}(t) = -2x(t), \end{cases}$$

and therefore,

$$\ddot{x}(t) + 4x(t) = 0.$$

As a consequence, $x(t)$ is of the form $x(t) = \lambda\cos 2t + \mu\sin 2t$ with $\lambda, \mu \in \mathbf{R}^n$, and returning to the system of Hamilton–Jacobi, this also gives $\xi(t) = -\lambda\sin 2t + \mu\cos 2t$. Writing down the initial conditions $x(0) = x_0$ and $\xi(0) = \xi_0$, we finally obtain

$$\exp tH_{x^2+\xi^2}(x_0,\xi_0) = (x_0\cos 2t + \xi_0\sin 2t, -x_0\sin 2t + \xi_0\cos 2t).$$

One can observe that when t describes $\mathbf{R}$, then $\exp tH_{x^2+\xi^2}(x_0,\xi_0)$ describes the circle of radius $(x_0^2 + \xi_0^2)^{1/2}$ centered at the origin and contained in the plane generated by x_0 and ξ_0.

Remark 4.3.3 : **Stationary Points** If (x_0, ξ_0) satisfies $(\nabla_{x,\xi}p)(x_0,\xi_0) = 0$, then for all $t \in \mathbf{R}$ one has $\exp tH_p(x_0,\xi_0) = (x_0,\xi_0)$. This is just a consequence of the unicity of the solution of the system of Hamilton–Jacobi, and the fact that $t \mapsto (x_0, \xi_0)$ is such a solution.

Remark 4.3.4 There are cases where $\exp tH_p(x_0, \xi_0)$ is not defined for all $t \in \mathbf{R}$. As an example, one can take $p(x, \xi) = x^2\xi$ on $\mathbf{R}^2$. Then one obtains

$$\exp tH_p(x_0, \xi_0) = \left(\frac{x_0}{1 - tx_0}, \xi_0(1 - tx_0)^2\right),$$

which, if $x_0 \neq 0$, becomes singular as t approaches $1/x_0$. In the case of the symbol of the Schrödinger operator $p(x, \xi) = \xi^2 + V(x)$, the same phenomenon happens, e.g., for $V(x) = -x^4/2$ (take $x(t) = (t-1)^{-1}$), but one can prove that if V is uniformly bounded from below, then $\exp tH_p(x_0, \xi_0)$ is always defined for all $t \in \mathbf{R}$ (see Exercise 5 at the end of this chapter).

The main basic properties of the Hamilton flow are described in the following proposition:

Proposition 4.3.5 *For all* $(x_0, \xi_0) \in \mathbf{R}^{2n}$ *one has*
(i) $\forall\, t, s \in \mathbf{R}$ *small enough,*

$$(\exp(t + s) H_p)(x_0, \xi_0) = (\exp tH_p)(\exp sH_p(x_0, \xi_0)).$$

(In particular, $\exp(-tH_p) \circ \exp tH_p = I_{\mathbf{R}^{2n}}$*.)*
(ii) $\forall t \in \mathbf{R}$ *small enough,*

$$p(\exp tH_p(x_0, \xi_0)) = p(x_0, \xi_0).$$

Proof The identity (i) is an immediate consequence of the fact that the two members solve the same first-order differential equation with respect to t, and have the same value at $t = 0$. For (ii), we obtain by differentiating with respect to t,

$$\begin{aligned}\frac{\partial}{\partial t} p(\exp tH_p(x_0, \xi_0)) &= (\nabla p)(\exp tH_p(x_0, \xi_0)) \cdot H_p(\exp tH_p(x_0, \xi_0)) \\ &= \left[\frac{\partial p}{\partial x} \cdot \frac{\partial p}{\partial \xi} + \frac{\partial p}{\partial \xi} \cdot \left(-\frac{\partial p}{\partial x}\right)\right] (\exp tH_p(x_0, \xi_0)) = 0,\end{aligned}$$

which means that the function $t \mapsto p(\exp tH_p(x_0, \xi_0))$ is constant, and therefore equals its value at $t = 0$, which is $p(x_0, \xi_0)$. ◇

Remark 4.3.6 In particular, if $(x_0, \xi_0) \in p^{-1}(0)$, then for all $t \in \mathbf{R}$ one has $\exp tH_p(x_0, \xi_0) \in p^{-1}(0)$. In other words, the hypersurface $\{p(x, \xi) = 0\}$ is stable under the action of the flow of H_p. By definition, any curve of the type $\{\exp tH_p(x_0, \xi_0)\ ;\ t \in]T_0, T_1[\}$ with $p(x_0, \xi_0) = 0$ is called a *null-bicharacteristic* of p.

(b) Propagation

Now we come to the main argument of this section, which consists in the relationship between the microsupport of a solution of a PDE, and the Hamilton flow associated with the equation. Indeed, as we shall see hereinafter, in this case the microsupport of the solution appears to be invariant by the Hamilton flow. Such a property is called *propagation* of the microsupport, since it can be expressed by saying that any point in the microsupport gives rise to a whole curve passing through this point and contained in the microsupport (at least if such a point is not stationary in the sense of Remark 4.3.3).

Let $m \in \mathbf{R}$ and let $p = p(x,\xi,h) \in S_{2n}^{hol}(\langle\xi\rangle^m)$ be of the form

$$p(x,\xi;h) = p_0(x,\xi) + \varepsilon(h)r(x,\xi;h) \tag{4.3.3}$$

with $p_0 \in S_{2n}^{hol}(\langle\xi\rangle^m)$ independent of h, $r \in S_{2n}^{hol}(\langle\xi\rangle^m)$, $\varepsilon(h) \in \mathbf{C}$ tending to 0 as $h \to 0$, and such that

$$p_0(x,\xi) \in \mathbf{R} \text{ for all } (x,\xi) \in \mathbf{R}^{2n}. \tag{4.3.4}$$

We investigate the solutions $u \in L^2(\mathbf{R}^n)$ of the system

$$\begin{cases} Pu = 0, \\ \|u\|_{L^2} \le 1, \end{cases} \tag{4.3.5}$$

where $P = \mathrm{Op}_h^t(p)$ ($t \in [0,1]$ fixed arbitrarily). We have the following theorem:

Theorem 4.3.7 (Propagation of the Microsupport) *Assume (4.3.3) and (4.3.4), and let $u \in L^2(\mathbf{R}^n)$ satisfying (4.3.5). Let also $(x_0,\xi_0) \in \mathbf{R}^{2n}$ and assume that $\exp tH_{p_0}(x_0,\xi_0)$ exists for $t \in]T_0,T_1[$, $T_0 < 0 < T_1$. Then one has the equivalences*

$$\begin{aligned} (x_0,\xi_0) \in \mathrm{MS}(u) \quad &\Leftrightarrow \quad \forall t \in]T_0,T_1[, \quad \exp tH_{p_0}(x_0,\xi_0) \in \mathrm{MS}(u) \\ &\Leftrightarrow \quad \exists t \in]T_0,T_1[, \quad \exp tH_{p_0}(x_0,\xi_0) \in \mathrm{MS}(u). \end{aligned}$$

(In other words, $\mathrm{MS}(u)$ is invariant under the flow of H_{p_0}.)

Proof By Proposition 4.3.5(ii) and Theorem 4.2.2, we can assume without loss of generality that $p_0(x_0,\xi_0) = 0$. We can also restrict ourselves to the case $m = 0$ by using the following result:

Lemma 4.3.8 *Assume* $p_0(x_0, \xi_0) = 0$ *and let* $q \in C^\infty(\mathbf{R}^{2n})$ *real-valued and nonvanishing. Then for all* $t \in]T_0, T_1[$, *there exists* $s_t \in \mathbf{R}$ *such that*

$$\exp tH_{p_0}(x_0, \xi_0) = \exp s_t H_{qp_0}(x_0, \xi_0).$$

Moreover, there exists $S_0 < 0 < S_1$ *such that the application* $t \mapsto s_t$ *is a* C^∞ *diffeomorphism from* $]T_0, T_1[$ *to* $]S_0, S_1[$.

Proof of the Lemma Set $b = \dfrac{1}{q}$ ($\in C^\infty(\mathbf{R}^{2n})$ real-valued) and $a = qp_0$. Since $p_0 = ab$, we have

$$(ab)(\exp tH_{p_0}(x_0, \xi_0)) = 0,$$

and therefore, since b never vanishes,

$$a(\exp tH_{p_0}(x_0, \xi_0)) = 0$$

for all $t \in]T_o, T_1[$. As a consequence, writing $(x(t), \xi(t)) = \exp tH_{p_0}(x_0, \xi_0)$, we get

$$\begin{cases} \dot{x}(t) = \dfrac{\partial p_0}{\partial \xi} = a(x(t), \xi(t))\dfrac{\partial b}{\partial \xi} + b\dfrac{\partial a}{\partial \xi} = b\dfrac{\partial a}{\partial \xi}(x(t), \xi(t)) \\ \dot{\xi}(t) = -a\dfrac{\partial b}{\partial x} - b\dfrac{\partial a}{\partial x} = -b\dfrac{\partial a}{\partial x}(x(t), \xi(t)) \end{cases}$$

that is, denoting $f(t) = b(x(t), \xi(t))$:

$$\begin{cases} \dot{x}(t) = f(t)\dfrac{\partial a}{\partial \xi}(x(t), \xi(t)), \\ \\ \dot{\xi}(t) = -f(t)\dfrac{\partial a}{\partial x}(x(t), \xi(t)), \end{cases}$$

with, moreover, $x(0) = x_0$ and $\xi(0) = \xi_0$.

Now, let $(y(t), \eta(t)) = \exp tH_a(x_0, \xi_0)$, and define

$$s(t) = \int_0^t f(s)ds.$$

Since f is real-valued and $\dot{s}(t) = f(t)$ never vanishes, we see that the map $t \mapsto s(t)$ is a C^∞ diffeomorphism from $]T_0, T_1[$ to some interval $]S_0, S_1[$ (also with $S_0 < 0 < S_1$, since s(0)=0), and we have

$$\begin{cases} \dfrac{\partial}{\partial t}(y(s(t)) = f(t)\dfrac{\partial a}{\partial \xi}(y(s(t)), \eta(s(t)), \\ \\ \dfrac{\partial}{\partial t}(\eta(s(t)) = -f(t)\dfrac{\partial a}{\partial x}(y(s(t)), \eta(s(t)), \end{cases}$$

with, moreover, $y(s(0)) = x_0$ and $\eta(s(0)) = \xi_0$. In particular, $(x(t), \xi(t))$ and $(y(s(t), \eta(s(t)))$ solve the same first-order differential system with respect to t and have the same initial data. They must therefore be equal; that is,

$$\exp tH_{p_0}(x_0, \xi_0) = \exp s(t) H_a(x_0, \xi_0)$$

for all $t \in]T_0, T_1[$. ◇

Using Lemma 4.3.8 with $q = \langle\xi\rangle^{-m}$, we see that we can work with the operator $\mathrm{Op}_h^t(\langle\xi\rangle^{-m})\mathrm{Op}_h^t(p)$ instead of $\mathrm{Op}_h^t(p)$, that is, (thanks to the symbolic calculus) we can assume that p_0 and r are in $S_{2n}^{hol}(1)$.

Now we claim that it is enough to prove the following assertion:

$$\left\{ \begin{array}{l} \text{For all } (x_0, \xi_0) \in \mathrm{MS}(u), \text{ there exists } \delta > 0 \text{ such that} \\ \exp tH_{p_0}(x_0, \xi_0) \in \mathrm{MS}(u) \text{ for all } t \in [-\delta, \delta]. \end{array} \right. \tag{4.3.6}$$

Indeed, if (4.3.6) is true, then the set $\{t \in]T_0, T_1[\ ;\ \exp tH_{p_0}(x_0, \xi_0) \in \mathrm{MS}(u)\}$ will be both open and closed in $]T_0, T_1[$, and therefore equal to $]T_0, T_1[$, since this latter is connected.

First of all, notice that (4.3.6) is obvious if $\nabla_{x,\xi} p_0(x_0, \xi_0) = 0$, since in this case $H_{p_0}(x_0, \xi_0) = 0$ and thus $\exp tH_{p_0}(x_0, \xi_0) = (x_0, \xi_0)$ for all $t \in \mathbf{R}$. We can therefore assume from now on that $\nabla_{x,\xi} p_0(x_0, \xi_0) \neq 0$.

Let us prove (4.3.6) by contradiction; i.e., assume that for all $\delta > 0$ there exists $t_\delta \in [-\delta, \delta]$ such that

$$\exp t_\delta H_{p_0}(x_0, \xi_0) \notin \mathrm{MS}(u).$$

Setting $(x_\delta, \xi_\delta) = \exp t_\delta H_{p_0}(x_0, \xi_0)$, this means that there exist $\alpha_\delta > 0$ and a neighborhood W_δ of (x_δ, ξ_δ) such that

$$Tu(x, \xi; h) = \mathcal{O}(e^{-\alpha_\delta/h}) \ \text{ for } (x, \xi) \in W_\delta. \tag{4.3.7}$$

To have a contradiction, it is enough to prove that (4.3.7) implies that $(x_0, \xi_0) \notin \mathrm{MS}(u)$. We first prove the following lemma:

Lemma 4.3.9 *Assume (4.3.4) and (4.3.5) with $m = 0$, and let $P = \mathrm{Op}_h^t(p)$ ($t \in [0, 1]$ arbitrary). Then for all $g = g(x, \xi) \in S_{2n}(1)$ real-valued, there exists $C > 0$ such that for all $\theta \in \mathbf{R}_+$ and for all $v \in L^2(\mathbf{R}^n)$, one has*

$$\left\| e^{\theta g/h} TPv \right\|^2 \geq \theta^2 \left\| (H_{p_0} g) e^{\theta g/h} Tv \right\|^2 - C(h + |\varepsilon(h)| + \theta^3) \left\| e^{\theta g/h} Tv \right\|^2$$

uniformly with respect to $\theta, h > 0$ both small enough, and $v \in L^2(\mathbf{R}^n)$.

Proof We apply Corollary 3.5.5 with $\psi = \theta g$, which gives

$$\left\| e^{\theta g/h} T \mathrm{Op}_h^t(p_0) v \right\|^2 = \left\| p_0(x - 2\theta\partial_z g, \xi + 2i\theta\partial_z g) e^{\theta g/h} T v \right\|^2 + \mathcal{O}(h) \left\| e^{\theta g/h} T v \right\|^2$$

(where as usual, $\partial_z = (\partial_x + i\partial_\xi)/2$), and therefore, by the triangle inequality and the Calderón–Vaillancourt theorem,

$$\left\| e^{\theta g/h} T P v \right\|^2 = \left\| p_0(x - 2\theta\partial_z g, \xi + 2i\theta\partial_z g) e^{\theta g/h} T v \right\|^2 + \mathcal{O}(h + |\varepsilon(h)|) \left\| e^{\theta g/h} T v \right\|^2. \tag{4.3.8}$$

Now, taking the Taylor expansion near 0 with respect to θ, we get

$$p_0(x - 2\theta\partial_z g, \xi + 2i\theta\partial_z g) = p_0(x,\xi) - 2\theta \frac{\partial p_0}{\partial x}\partial_z g + 2i\theta \frac{\partial p_0}{\partial \xi}\partial_z g + \mathcal{O}(\theta^2),$$

and thus, since p_0 is real on the real space $\mathbf{R}^{2n}$,

$$\begin{aligned} \operatorname{Im} p_0(x - 2\theta\partial g, \xi + 2i\theta\partial g) &= -\theta \frac{\partial p_0}{\partial x}\frac{\partial g}{\partial \xi} + \theta \frac{\partial p_0}{\partial \xi}\frac{\partial g}{\partial x} + \mathcal{O}(\theta^2) \\ &= \theta H_{p_0} g + \mathcal{O}(\theta^2) \end{aligned} \tag{4.3.9}$$

in $S_{2n}(1)$. Inserting (4.3.9) into (4.3.8), the lemma follows by a new application of the triangle inequality and the Calderón–Vaillancourt theorem. ◇

Now we are going to apply Lemma 4.3.9 to some convenient g, which in some sense will permit us to "slide" along the bicharacteristic of p_0 passing through (x_0, ξ_0).

By assumption, there exists a neighborhood V of (x_0, ξ_0) such that $\nabla_{x,\xi} p_0$ never vanishes on V. Then also H_{p_0} never vanishes on V, and possibly by taking V smaller, we see that there exist local coordinates $y = (y_1, \dots, y_{2n})$ on V, centered at (x_0, ξ_0), and such that in these coordinates H_{p_0} becomes

$$H_{p_0} = \frac{\partial}{\partial y_1}.$$

Taking $\delta > 0$ small enough, one can also assume that $\overline{W_\delta} \subset V$, and then the point (x_δ, ξ_δ) becomes $(t_\delta, 0)$ in these coordinates.

Now choose $a, b \in \mathbf{R}$ and $c > 0$ (all of them depending on δ) such that $a < t_\delta < b$ and

$$\{(y,\eta)\ ;\ a \leq y_1 \leq b\,,\ |y'| \leq c\} \subset W_\delta, \tag{4.3.10}$$

where we have used the notation $y = (y_1, y') \in \mathbf{R} \times \mathbf{R}^{2n-1}$ for $y \in \mathbf{R}^{2n}$.

Then we consider a function $f = f(y_1) \in C^\infty(\mathbf{R})$ satisfying, for some $\beta, d > 0$ small enough,

$$\begin{cases} f = 0 \text{ outside } [-2d, b]; \\ 0 \leq f \leq \alpha_\delta \text{ everywhere}; \\ f' \leq -\beta \text{ on } [-d, a]; \\ f' \leq 0 \text{ on } [-2d, -d]; \\ f(0) = \dfrac{\alpha_\delta}{2}. \end{cases} \tag{4.3.11}$$

We also consider a cutoff function $\chi \in C_0^\infty(]-c, c[\ ;\ [0,1])$ satisfying

$$\begin{cases} \chi = 1 \text{ on } \left[-\dfrac{c}{4}, \dfrac{c}{4}\right]; \\ \chi \neq 0 \text{ on } \left[-\dfrac{c}{2}, \dfrac{c}{2}\right]; \\ \chi \leq \dfrac{1}{4} \text{ outside } \left[-\dfrac{c}{2}, \dfrac{c}{2}\right]. \end{cases} \tag{4.3.12}$$

Then we set

$$g(y) = \chi(|y'|) f(y_1),$$

which can be extended to a smooth function on $\mathbf{R}^{2n}$ by taking 0 outside V. Then since $H_{p_0} g = \chi(|y'|) f'(y_1)$ on V, we have by (4.3.11) and (4.3.12),

$$\begin{cases} H_{p_0} g = 0 \text{ outside } V; \\ H_{p_0} g < 0 \text{ on } V_\delta := \left\{(y, \eta)\ ;\ y_1 \in [-d, a], |y'| \leq \dfrac{c}{2}\right\}; \\ g(x_0, \xi_0) = \dfrac{\alpha_\delta}{2}; \\ g \leq \dfrac{\alpha_\delta}{4} \text{ for } |y'| + |\eta| \geq \dfrac{c}{2}; \\ g \leq \alpha_\delta \text{ everywhere}. \end{cases} \tag{4.3.13}$$

Now, Lemma 4.3.9 applied to $v = u$ implies

$$\left\|(H_{p_0}g)e^{\theta g/h}Tu\right\|^2_{L^2(V_\delta)} \leq C\left(\frac{h+|\varepsilon(h)|}{\theta^2}+\theta\right)\left\|e^{\theta g/h}Tu\right\|^2, \tag{4.3.14}$$

where $C > 0$ depends on g (and therefore on δ). Moreover, by (4.3.13) we see that there exists $C_\delta > 0$ such that $|H_{p_0}g|^2 \geq \dfrac{1}{C_\delta}$ on V_δ . As a consequence, we get from (4.3.14),

$$\left\|e^{\theta g/h}Tu\right\|^2_{L^2(V_\delta)} \leq CC_\delta\left(\frac{h+|\varepsilon(h)|}{\theta^2}+\theta\right)\left\|e^{\theta g/h}Tu\right\|^2,$$

and therefore, fixing θ small enough (depending on δ), we get for all $h > 0$ small enough,

$$\left\|e^{\theta g/h}Tu\right\|^2_{L^2(V_\delta)} \leq C'\left\|e^{\theta g/h}Tu\right\|^2_{L^2(\mathbf{R}^{2n}\backslash V_\delta)} \tag{4.3.15}$$

for some other positive constant $C' = C'(\delta)$.

Now we notice that by (4.3.7), (4.3.10), (4.3.12), and (4.3.13),

- On $\mathbf{R}^{2n}\backslash V$: $g = 0$
- On $V\backslash V_\delta$:

  If $y_1 \in [a, b]$ and $|y'| \leq \frac{c}{2}$, then $e^{\theta g/h}Tu = \mathcal{O}(e^{(\theta-1)\alpha_\delta/h})$;

  If $|y'| \geq \frac{c}{2}$, then $g \leq \dfrac{\alpha_\delta}{4}$;

  If $y_1 \geq b$, then $g = 0$;

  If $y_1 \leq -d$, then $g \leq \dfrac{\alpha_\delta}{2} - \beta d$.

As a consequence (since we can also assume that $\theta \leq 1$ and $\beta d \leq \alpha_\delta/4$),

$$\left\|e^{\theta g/h}Tu\right\|_{L^2(\mathbf{R}^{2n}\backslash V_\delta)} = \mathcal{O}\left(1+e^{\theta(\frac{\alpha}{2}-\beta d)/h}\right).$$

Inserting this estimate into (4.3.15), we get

$$\left\|e^{\theta g/h}Tu\right\|_{L^2(V_\delta)} = \mathcal{O}\left(1+e^{\theta(\frac{\alpha_\delta}{2}-\beta d)/h}\right), \tag{4.3.16}$$

and therefore, if $V_\delta' \subset V_\delta$ is a small enough neighborhood of (x_0, ξ_0) such that $g\big|_{V_\delta'} \geq \frac{\alpha_\delta}{2} - \frac{\beta d}{2}$ (such a neighborhood exists, since $g(x_0, \xi_0) = \frac{\alpha_\delta}{2}$), we deduce from (4.3.16) that

$$\|Tu\|_{L^2(V_\delta')} = \mathcal{O}\left(e^{\theta(\frac{\alpha}{2}-\beta d)/h-\theta(\frac{\alpha}{2}-\frac{\beta d}{2}/h)}\right) = \mathcal{O}\left(e^{-\theta\beta d/2h}\right). \tag{4.3.17}$$

This implies that $(x_0, \xi_0) \notin \mathrm{MS}(u)$, which contradicts the assumption we started from. This finishes the proof of Theorem 4.3.7. ◇

Remark 4.3.10 In fact, we see from the proof that the assumption $Pu = 0$ can be replaced by $TPu = \mathcal{O}(e^{-\alpha/h})$ near $\{\exp tH_p(x_0, \xi_0)\ ;\ t \in]T_0, T_1[\}$ (for some $\alpha > 0$), i.e., $\exp tH_p(x_0, \xi_0) \notin \mathrm{MS}(Pu)$ for $t \in]T_0, T_1[$.

Remark 4.3.11 Theorem 4.3.7 is interesting only for those points (x_0, ξ_0) satisfying $\nabla_{x,\xi} p_0(x_0, \xi_0) \neq 0$. Then one says that (x_0, ξ_0) is a *simple characteristic point* of P, or also that P is of *principal type* at (x_0, ξ_0). If this is not the case, one may try to use instead Theorem 4.4.3 below.

Remark 4.3.12 If p_0 is not assumed to be real on the real space, then the result of Theorem 4.3.7 remains valid, provided that the bicharacteristic $\gamma := \{\exp tH_p(x_0, \xi_0)\ ;\ t \in]T_0, T_1[\}$ is real. This is the so-called *Hanges theorem*: see Exercise 6 at the end of this chapter.

Remark 4.3.13 If instead of $\mathrm{MS}(u)$ one investigates $\mathrm{FS}(u)$ (see Section 2.8), then the same result of propagation is valid, even without the assumption of analyticity on p, see Exercise 7 of this chapter.

Remark 4.3.14 Here we have made a global assumption on u, namely, that $u \in L^2(\mathbf{R}^n)$. Actually, in the case where P is a *differential* operator, it is enough to replace it by the following *local* assumption: $u \in H^s_{loc}(\Omega)$, where $s \in \mathbf{R}$ is arbitrary, and Ω is an open neighborhood of $\{\Pi_x(\exp tH_p(x_0, \xi_0))\ ;\ t \in]T_0, T_1[\}$. Indeed, working with $\langle hD_x\rangle^s \chi u$ instead of u, where $\chi \in C_0^\infty(\Omega)$, the proof can be adapted by taking g such that $\chi = 1$ near its support (see exercise 8 at the end of this chapter).

Application

In the case of a normalized eigenfunction of the Schrödinger operator,

$$\begin{cases} Hu = Eu, \\ \|u\|_{L^2} = 1, \end{cases} \tag{4.3.18}$$

where $H = -h^2\Delta + V$ (with $V \in S_n^{hol}(1)$), and $E = E(h) \to E_0$ in $\mathbf{R}$ as $h \to 0$, then the null bicharacteristics of $P = H - E$ are given by

$$\begin{cases} \dot{x}(t) = 2\xi(t), \\ \dot{\xi}(t) = -\nabla V(x(t)), \\ \xi(0)^2 + V(x(0)) = E_0, \end{cases} \tag{4.3.19}$$

and one recognizes the *classical trajectories of energy* E_0 (see the introduction). Therefore, in this case Theorem 4.3.7 shows that $\mathrm{MS}(u)$ is a union of maximal classical trajectories of energy E_0.

4.4 Microhyperbolicity

In the previous section we have shown that for a solution of a PDE, the points of its microsupport that are *not stationary* give rise to a whole curve also contained in the microsupport. Now we investigate those points of the microsupport that are stationary, that is, points (x_0, ξ_0) at which both p_0 and H_{p_0} vanish (where p_0 is the principal symbol of the equation). As we have already noticed, the result of the previous section is irrelevant for such points.

Therefore, we again consider a function $u \in L^2(\mathbf{R}^n)$ satisfying

$$\begin{cases} Pu = 0 \\ \|u\|_{L^2} \leq 1 \end{cases} \tag{4.4.1}$$

where $P = \mathrm{Op}_h^t(p)$ with $t \in [0, 1]$ fixed, and $p = p(x, \xi, h) \in S_{2n}^{hol}(\langle\xi\rangle^m)$ of the form

$$p(x, \xi; h) = p_0(x, \xi) + \varepsilon(h) r(x, \xi; h) \tag{4.4.2}$$

with p_0 independent of h and $\varepsilon(h) \in \mathbf{C}$ tending to 0 as $h \to 0$. (Here we do not assume that p_0 is real on the real space.)

The result of propagation we are going to prove relies on the behavior of p_0 in the complex space, and more precisely on a property called *microhyperbolicity* (in some direction), which is defined as follows:

Definition 4.4.1 *Let $(x_0, \xi_0) \in \mathbf{R}^{2n}$ and let ϕ be a real C^∞ function defined near (x_0, ξ_0). One says that $P = \mathrm{Op}_h^t(p_0 + \varepsilon(h)r)$ is* **microhyperbolic** *at (x_0, ξ_0) in the direction H_ϕ if there exist a real neighborhood U of (x_0, ξ_0) and a constant $\delta_0 > 0$ such that for all (x, ξ) in U and for all $\delta \in (0, \delta_0]$, one has*

$$p_0\left(x + i\delta\frac{\partial\phi}{\partial\xi}(x_0, \xi_0), \xi - i\delta\frac{\partial\phi}{\partial x}(x_0, \xi_0)\right) \neq 0. \tag{4.4.3}$$

Remark 4.4.2 By taking the Taylor expansion with respect to δ, we see that

$$p_0(x+i\delta\partial_\xi\phi(x_0, \xi_0), \xi-i\delta\partial_x\phi(x_0, \xi_0)) = p(x, \xi)+i\delta H_{p_0}\phi(x_0, \xi_0)+\mathcal{O}(\delta^2) \tag{4.4.4}$$

in $S_{2n}(\langle\xi\rangle^m)$. As a consequence:

- If P is elliptic at (x_0, ξ_0), then it is automatically microhyperbolic at (x_0, ξ_0) in any direction;

- If p_0 is real on the real space and $(x_0, \xi_0) \in \mathrm{Char}(P)$ is not a stationary point for H_{p_0}, then P is microhyperbolic at (x_0, ξ_0) in the direction H_ϕ for all ϕ such that $H_{p_0}\phi(x_0, \xi_0) \neq 0$ (that is, such that the hypersurface of equation $\phi(x, \xi) = \phi(x_0, \xi_0)$ is transversal to H_{p_0} at (x_0, ξ_0)).

But as we shall see, the result we are going to prove is interesting (that is, says more than we already know) only at those points $(x_0, \xi_0) \in \mathrm{Char}(P)$ that are *stationary* for H_{p_0}.

Example 1 In the case $n = 1$, take $p_0 = x\xi$ and $x_0 = \xi_0 = 0$ (so that both p_0 and H_{p_0} vanish there). Then one can see that for all $r \in S_2(\langle x\rangle\langle\xi\rangle)$, the operator $P = \mathrm{Op}_h^0(p_0 + hr) = xhD_x + h\mathrm{Op}_h^0(r)$ is microhyperbolic at $(0, 0)$ in the direction $H_{\xi+\mu x}$ for all $\mu \in \mathbf{R}^*$. Indeed, the quantity that one has to investigate becomes here

$$(x + i\delta)(\xi - i\mu\delta) = x\xi + \mu\delta^2 + i\delta(\xi - \mu x),$$

which never vanishes for $\delta \neq 0$ and $(x, \xi) \in \mathbf{R}^2$.

Example 2 In the same spirit (but to remain even closer to quantum mechanics and Schrödinger operators), if $P = h^2D_x^2 + V(x)$ on $L^2(\mathbf{R})$ with $V \in S_1(1)$, $V(x) = -x^2+\mathcal{O}(x^3)$ near 0, then P is microhyperbolic at $(0, 0)$ in the directions H_x and $H_{\xi+\nu x}$ for all $\nu \neq \pm 1$.

Example 3 An example in higher dimension can be easily obtained by perturbing the previous examples, e.g., $p(x,\xi) = x_n\xi_n + x_n^2 a(x',\xi')$ with $a(0,0) \geq 0$ and $a(0,0) = 0$, which is microhyperbolic at $(0,0)$ in the direction $H_{\xi_n+\mu x_n}$ for all $\mu > 0$.

The result of propagation for microhyperbolic analytic operators (which is due to Kawai and Kashiwara [KK]) is the following one:

Theorem 4.4.3 (Kawai–Kashiwara) *Assume (4.4.2) with p_0 and r in $S_{2n}^{hol}(\langle\xi\rangle^m)$, p_0 independent of h, and $\varepsilon(h) \to 0$ as $h \to 0$. Let $u \in L^2(\mathbf{R}^n)$ be a solution of (4.4.1), and assume also that P is microhyperbolic at some point $(x_0,\xi_0) \in \mathbf{R}^{2n}$ in some direction H_ϕ, and that there exists a real neighborhood U of (x_0,ξ_0) such that*

$$\mathrm{MS}(u) \cap \{(x,\xi) \in U \ ; \ \phi(x,\xi) < \phi(x_0,\xi_0)\} = \emptyset.$$

Then

$$(x_0,\xi_0) \notin \mathrm{MS}(u).$$

Proof The idea of the proof is taken from [Sj1], although our use of Corollary 3.5.5 makes it considerably simpler.

First of all, composing P to its left by $\mathrm{Op}_h^t(\langle\xi\rangle^{-m})$, one can assume without loss of generality that $m = 0$.

Moreover, by a standard result of analytic function theory (called the *Bochner tube theorem*, see, e.g., [Ho1] vol. 1 Section 8.7), the property (4.4.3) enjoys some stability with respect to $H_\phi(x_0,\xi_0)$, in the sense that if P is microhyperbolic at (x_0,ξ_0) in the direction H_ϕ, then there exists a real neighborhood U of (x_0,ξ_0) *and* a real neighborhood V of $H_\phi(x_0,\xi_0)$ such that for all $(x,\xi) \in U$, $(x^*,\xi^*) \in V$, and $\delta > 0$ small enough, one has

$$p_0(x + i\delta x^*, \xi + i\delta\xi^*) \neq 0. \tag{4.4.5}$$

Now let $r > 0$ be sufficiently small so that the ball $B_{2r} \subset \mathbf{R}^{2n}$ centered at (x_0,ξ_0) of radius $2r$ is included in U and $\phi \in C^\infty(B_{2r})$, and let $\chi \in C_0^\infty(B_{2r} \ ; \ [0,1])$ such that $\chi = 1$ on B_r. We set

$$\phi_r(x,\xi) = \chi(x,\xi)\left(\phi(x,\xi) - \phi(x_0,\xi_0) - \frac{r^3}{2} + r(x-x_0)^2 + r(\xi-\xi_0)^2\right).$$

Then extending ϕ_r by 0 outside B_{2r}, we have

$$\begin{cases} \phi_r \in C_0^\infty(\mathbf{R}^{2n} \,;\, \mathbf{R}); \\ H_{\phi_r}(x,\xi) = H_\phi(x,\xi) + 2r(\xi - \xi_0) - 2r(x - x_0) \;\text{ for } (x,\xi) \in B_r; \\ \phi_r(x_0,\xi_0) = -\dfrac{r^3}{2} < 0 \\ \phi_r \geq \chi \cdot \left(\phi - \phi(x_0,\xi_0) + \dfrac{r^3}{2}\right) \;\text{ outside } B_r. \end{cases} \tag{4.4.6}$$

In particular, possibly by shrinking r, one can also assume that

$$H_{\phi_r}(x,\xi) \in V \;\text{ for } (x,\xi) \in B_r. \tag{4.4.7}$$

Now we apply Corollary 3.5.5 with $f = 1$ and $\psi = -\delta\phi_r$, where $\delta > 0$ will be taken small enough later. We get

$$\begin{aligned} \left\| e^{-\delta\phi_r/h} T P u \right\|^2 &= \left\| p_0(x + 2\delta\partial_z\phi_r, \xi - 2i\delta\partial_z\phi_r) e^{-\delta\phi_r/h} T u \right\|^2 \\ &\quad + \mathcal{O}(h + |\varepsilon(h)|) \left\| e^{-\delta\phi_r/h} T u \right\|^2, \end{aligned}$$

and therefore, using (4.4.1),

$$\begin{aligned} &\left\| p_0(x + \delta\partial_x\phi_r + i\delta\partial_\xi\phi_r, \xi + \delta\partial_\xi\phi_r - i\delta\partial_x\phi_r) e^{-\delta\phi_r/h} T u \right\|^2 \\ &= \mathcal{O}(h + |\varepsilon(h)|) \left\| e^{-\delta\phi_r/h} T u \right\|^2. \end{aligned}$$

In particular,

$$\begin{aligned} &\left\| p_0(x + \delta\partial_x\phi_r + i\delta\partial_\xi\phi_r, \xi + \delta\partial_\xi\phi_r - i\delta\partial_x\phi_r) e^{-\delta\phi_r/h} T u \right\|^2_{L^2(B_r)} \\ &= \mathcal{O}(h + |\varepsilon(h)|) \left\| e^{-\delta\phi_r/h} T u \right\|^2, \end{aligned} \tag{4.4.8}$$

and if δ is chosen small enough, one also has

$$(x + \delta\partial_x\phi_r(x,\xi), \xi + \delta\partial_\xi\phi_r(x,\xi)) \in U \;\text{ for } (x,\xi) \in B_r.$$

As a consequence, using (4.4.5) and (4.4.7), we see that for $\delta > 0$ small enough there exists a constant $C_\delta > 0$ such that

$$|p_0(x + \delta\partial_x\phi_r + i\delta\partial_\xi\phi_r, \xi + \delta\partial_\xi\phi_r - i\delta\partial_x\phi_r)| \geq \frac{1}{C_\delta} \;\text{ for } (x,\xi) \in B_r. \tag{4.4.9}$$

Inserting (4.4.9) into (4.4.8), we get in particular

$$\left\| e^{-\delta\phi_r/h} Tu \right\|^2_{L^2(B_r)} = \mathcal{O}(h + |\varepsilon(h)|) \left\| e^{-\delta\phi_r/h} Tu \right\|^2,$$

and therefore, for h small enough,

$$\left\| e^{-\delta\phi_r/h} Tu \right\|^2_{L^2(B_r)} = \mathcal{O}(h + |\varepsilon(h)|) \left\| e^{-\delta\phi_r/h} Tu \right\|^2_{L^2(\mathbf{R}^{2n}\backslash B_r)}. \tag{4.4.10}$$

Now, since $(\mathbf{R}^{2n}\backslash B_r) \cap \{\phi_r < 0\}$ is included in $U \cap \{\phi < \phi(x_0, \xi_0) - \frac{r^3}{2}\}$ (see (4.4.6)), the assumption on MS(u) implies that for $\delta > 0$ sufficiently small,

$$\left\| e^{-\delta\phi_r/h} Tu \right\|^2_{L^2((\mathbf{R}^{2n}\backslash B_r)\cap\{\phi_r<0\})} = \mathcal{O}(1),$$

while on the other hand, one obviously has

$$\left\| e^{-\delta\phi_r/h} Tu \right\|^2_{L^2((\mathbf{R}^{2n}\backslash B_r)\cap\{\phi_r\geq 0\})} \leq \|Tu\|_{L^2} \leq 1.$$

In particular, we deduce from (4.4.10) that

$$\left\| e^{-\delta\phi_r/h} Tu \right\|^2_{L^2(B_r)} = \mathcal{O}(1),$$

and since $-\delta\phi_r(x_0, \xi_0) = \delta\frac{r^3}{2} > 0$, this proves that $(x_0, \xi_0) \notin$ MS(u). ◇

Application

In the situation of the previous Example 2, let $u_h \in L^2(\mathbf{R})$ be such that $\|u_h\|_{L^2} = 1$ and

$$Pu_h = E(h)u_h$$

with $E(h) \to 0$ as $h \to 0_+$ (possibly for h belonging to a sequence of numbers only). Then by the previous result, we see that if $\{\xi < 0\} \cap \text{MS}(u_h) \cap U = \emptyset$ (where U is a neighborhood of $(0,0)$), one has $(0,0) \notin \text{MS}(u_h)$, and therefore (since $\{0\} \times \mathbf{R} \cap \{\xi^2 + V(x) = 0\} = \{(0,0)\}$), $0 \notin \text{Suppexp}\, u_h$, i.e., u_h is exponentially small near 0. The same is true if $\{\xi > 0\} \cap \text{MS}(u_h) \cap U = \emptyset$, or $\{x < 0\} \cap \text{MS}(u_h) \cap U = \emptyset$, or $\{x > 0\} \cap \text{MS}(u_h) \cap U = \emptyset$.

As a consequence, if $0 \in \text{Supp}\exp u_h$, then necessarily one either has

$$\{\xi = \text{sign}(x)\sqrt{-V(x)}\} \cap U \subset \text{MS}(u_h)$$

or

$$\{\xi = -\text{sign}(x)\sqrt{-V(x)}\} \cap U \subset \text{MS}(u_h)$$

for some neighborhood U of $(0,0)$ (where $\text{sign}(x) = \pm 1$ denotes the sign of x).

Actually, such a result is also a consequence of the so-called *exact WKB constructions*, which is a technique proper to the one-dimensional case (see, e.g., [Vo]). However, by using the Kawai–Kashiwara theorem one can also get results in higher dimension, e.g., in the situation of the previous Example 3.

4.5 Boundary Value Problems

Now we investigate the solutions of boundary value problems, that is, problems of the form

$$\begin{cases} Pu = 0 \quad \text{in } \Omega, \\ B_1 u\,|_\Sigma = u_1\,, \\ \quad \vdots \\ B_m u\,|_\Sigma = u_m\,, \end{cases}$$

where P and $B_1, \dots, B_m$ are partial differential operators, Ω is an open set of $\mathbf{R}^n$, Σ is a hypersurface of Ω, and $u_1, \dots, u_m$ are given functions (e.g., in $L^2_{loc}(\Sigma)$). Since any solution u is usually defined only locally, we shall have to cut it off by multiplying it by a convenient smooth cutoff function.

The only kind of boundary value problem we consider here is the so-called *noncharacteristic Cauchy Problem*, which means that in convenient coordinates $x = (x_1, \dots, x_n)$ of $\mathbf{R}^n$ (centered near some a priori point near which one wants to study the solution), the boundary Σ is given by $\Sigma = \{x_n = 0\}$, the operators B_j ($j = 1, \dots, m$) are the first m normal derivatives to Σ,

$$B_j = \left(\frac{\partial}{\partial x_n}\right)^{j-1}, \qquad j = 1, \dots, m, \tag{4.5.1}$$

and the operator P is of the form

$$P = (hD_{x_n})^m + \sum_{j=0}^{m-1} A_j(x, hD_{x'})(hD_{x_n})^j \tag{4.5.2}$$

where each $A_j(x, hD_{x'})$ is an x_n-dependent partial differential operator with respect to $x' := (x_1, \dots, x_{n-1})$. of order at most $m - j$. More precisely, $A_j(x, hD_{x'})$ is of the form:

$$A_j(x, hD_{x'}) = \sum_{\substack{\alpha\in\mathbf{N}^{n-1}\\|\alpha|\leq m-j}} a_{j,\alpha}(x)(hD_{x'})^\alpha \tag{4.5.3}$$

with the $a_{j,\alpha}$'s analytic near $0 \in \mathbf{R}^n$.

For the sake of simplicity, we prefer to remain global with respect to the variables x', and in particular we assume that the coefficients $a_{j,\alpha}$ can be extended analytically globally with respect to x' (actually, this restriction could be removed by using more sophisticated arguments, such as those of [Sj1]). So, let us assume here that for some positive δ, and for all $j = 1, \dots, m-1$ and $|\alpha| \leq m - j$, one has

$$a_{j,\alpha} \text{ is holomorphic in } (\mathbf{R} + i]-\delta,\delta[)^{n-1} \times (]-\delta,\delta[+i]-\delta,\delta[). \tag{4.5.4}$$

Then one has the following version of the so-called *microlocal Holmgren theorem:*

Theorem 4.5.1 *Let P be the operator given in (4.5.2) with the conditions (4.5.3)-(4.5.4), and let $u \in C^\infty(]-\delta,\delta[_{x_n}; L^2(\mathbf{R}^{n-1}))$ satisfying*

$$\begin{cases} Pu = 0 \quad \textit{in } \mathbf{R}^{n-1}\times]-\delta,\delta[, \\ \|u(\cdot,x_n)\|_{L^2(\mathbf{R}^{n-1})} \leq 1 \quad \text{for all } x_n \in]-\delta,\delta[. \end{cases}$$

Then for all $(x_0', \xi_0') \in \mathbf{R}^{2(n-1)}$ such that

$$(x_0', \xi_0') \notin \mathrm{MS}(u\,|_{x_n=0}) \cup \mathrm{MS}(\partial_{x_n}u\,|_{x_n=0}) \cup \ldots \cup \mathrm{MS}(\partial_{x_n}^{m-1}u\,|_{x_n=0})$$

there exists $\delta' > 0$ such that for all $\chi \in C_0^\infty(]-\delta',\delta'[)$,

$$[\{x_0'\} \times \mathbf{R} \times \{\xi_0'\} \times \mathbf{R}] \cap \mathrm{MS}(\chi(x_n)u) = \emptyset.$$

Proof By setting $v = (u, hD_{x_n}u, \dots, (hD_{x_n})^{m-1}u)$, the equation becomes

$$hD_{x_n}v = A(x, hD_{x'})v, \tag{4.5.5}$$

where A is an $m \times m$ matrix of partial differential operators of order at most m. Let T' denote the partial FBI transform defined as in (3.1.1) but acting in the variables x' only, that is,

$$T'u(x, \xi'; h) = 2^{-\frac{n-1}{2}}(\pi h)^{-\frac{3(n-1)}{4}} \int e^{i(x'-y')\xi'/h-(x'-y')^2/2h}u(y', x_n)\,dy.$$

In particular, $T'v$ is a vectorial function of $(x', x_n, \xi'; h)$, and it is enough to prove that $T'v$ is exponentially small near $(x_0', 0, \xi_0')$ uniformly as h tends to 0.

Let $\psi = \psi(x', \xi') \in C_0^\infty(\mathbf{R}^{2(n-1)})$ be real-valued such that $\psi(x_0', \xi_0') > 0$ and

$$\left\| e^{\psi/h} T'v\,|_{x_n=0} \right\|_{L^2(\mathbf{R}^{2(n-1)})} = \mathcal{O}(1). \tag{4.5.6}$$

For any $\lambda > 0$, set $\psi_\lambda(x, \xi') = \psi(x', \xi') - \lambda |x_n|$ and set

$$g_\lambda(x_n) = \| \langle \xi' \rangle^{-m/2} e^{\psi_\lambda(x,\xi')/h} T'v \|^2_{L^2(\mathbf{R}^{2(n-1)})}.$$

Then, using Corollary 3.5.3, we get for $x_n \in [0, \delta)$,

$$\begin{aligned} hg_\lambda'(x_n) &= 2\mathrm{Re}\langle \frac{\widetilde{A}(x,\xi') - \lambda}{\langle \xi' \rangle^m} e^{\psi_\lambda(x,\xi')/h} T'v\ ,\ e^{\psi_\lambda(x,\xi')/h} T'v \rangle \\ &\quad + \mathcal{O}(h) \left\| e^{\psi_\lambda(x,\xi')/h} T'v \right\|^2 \end{aligned} \tag{4.5.7}$$

with $\widetilde{A}(x, \xi') = iA(x' - 2\partial_{z'}\psi, x_n, \xi' + 2i\partial_{z'}\psi) = \mathcal{O}(\langle \xi' \rangle^m)$ (here $z' = x' - i\xi'$). In particular, taking λ large enough so that (in the sense of $m \times m$ self-adjoint matrices) $\mathrm{Re}\widetilde{A}(x, \xi') - \lambda \leq -c_0 < 0$ on $\mathrm{Supp}\psi \times (-\delta, \delta)$, we get for $h > 0$ small enough,

$$hg_\lambda'(x_n) \leq C \|T'v\|^2_{L^2(\mathbf{R}^{2(n-1)})},$$

where C is a positive constant.Integrating from 0 to $x_n \geq 0$ and using the fact that $g_\lambda(0) = \mathcal{O}(1)$ uniformly as h tends to 0, this gives

$$g_\lambda(x_n) = \mathcal{O}(h^{-1}).$$

Using a similar argument for $x_n \leq 0$, we therefore get

$$\| \langle \xi' \rangle^{-m/2} e^{\psi(x',\xi')/h} T'v \|^2_{L^2(\mathbf{R}^{2(n-1)})} = \mathcal{O}(h^{-1} e^{2\lambda |x_n|/h})$$

uniformly with respect to $x_n \in (-\delta, \delta)$ and h small enough. Since $\psi(x_0', \xi_0') > 0$, the result follows for $|x_n| \leq \delta'$ by taking $\delta' > 0$ sufficiently small with respect to $\psi(x_0', \xi_0')/\lambda$. $\diamond$

4.6 Exercises and Problems

1. For $x \in \mathbf{R}$, let $u(x; h) = (\pi h)^{-1/2} e^{ix/h^2 - x^2/2h}$. Show that $\|u\|_{L^2} = 1$ and $0 \in \mathrm{Suppexp}u$, but $\{0\} \times \mathbf{R} \cap \mathrm{MS}(u) = \emptyset$. (Hint: One has $|Tu(x, \xi)| = (2\pi h)^{-1} e^{-x^2/4h - (1-h\xi)^2/4h^3}$.)

2. Show that the condition (4.1.6) is satisfied for those $u = u(x;h)$ that are analytic in an h-independent complex neighborhood of x_0 and verify there

$$|u(x)| \leq Ce^{C|\mathrm{Im}x|/h}$$

for some constant $C > 0$. (Hint: In the expression of Tu, make a change of contour of integration of the type $y \mapsto y - i\delta\chi(y)\xi/\langle\xi\rangle$, where $\delta > 0$ is small enough and χ is supported near x_0.)

3. **One-Dimensional Double Wells** - Let $V = V(x)$ be a holomorphic function in the complex strip $\mathcal{B} = \{x \in \mathbf{C}\ ;\ |\mathrm{Im}x| \leq a\}$ (with $a > 0$), such that for all $k \in \mathbf{N}$, $V^{(k)}$ is uniformly bounded in $\mathcal{B}$ and satisfies

$$V\,|_{\mathbf{R}} \geq 0;\ \ V^{-1}(0) = \{-1\ ;\ 1\};\ \ V'\big|_{]-\infty,-1[\cup]0,1[} < 0;$$
$$V'\big|_{]-1,0[\cup]1,+\infty[} > 0;\ \ \lim_{|x|\to+\infty} V(x) = E_1 > V(0)$$

For $h > 0$ small enough, we consider the Schrödinger operator

$$H = -h^2\frac{d^2}{dx^2} + V,$$

and we suppose there exist $E = E(h) \in \mathbf{R}$ and $u = u(x;h) \in L^2(\mathbf{R})$ such that

$$\begin{cases} Hu = Eu, \\ \|u\|_{L^2(\mathbf{R}^n)} = 1, \\ \lim_{h\to 0} E(h) = E_0 \in \mathbf{R}. \end{cases}$$

(1) Justify that

$$\mathrm{MS}(u) \subset \{(x,\xi) \in \mathbf{R}^2\ ;\ \xi^2 + V(x) = E_0\}.$$

(2) Prove the following assertions:

(a) If $E_0 \in]V(0), E_1[$, then $\mathrm{MS}(u) = \{(x,\xi) \in \mathbf{R}^2\ ;\ \xi^2 + V(x) = E_0\}$.

(b) If $E_0 = V(0)$, then necessarily $(0,0) \in \mathrm{MS}(u)$.

(c) If $E_0 \in [0, V(0)[$, then $\mathrm{MS}(u)$ contains at least one of the connected components of $\{(x,\xi) \in \mathbf{R}^2\ ;\ \xi^2 + V(x) = E_0\}$.

4. **Analytic Wave Front Set** - For $u \in \mathcal{S}'(\mathbf{R}^n)$ independent of h, one defines its *analytic wave front set* $\mathrm{WF}_a(u)$ by

$$\mathrm{WF}_a(u) := \mathrm{MS}(u) \cap \{\xi \neq 0\},$$

where $\mathrm{MS}(u)$ is as in Definition 3.2.1.

(i) Prove that $\mathrm{WF}_a(u)$ is a *conical* subset of $T^*\mathbf{R}^n\backslash 0$, in the sense that if $(x_0, \xi_0) \in \mathrm{WF}_a(u)$, then for all $\lambda > 0$, $(x_0, \lambda\xi_0) \in \mathrm{WF}_a(u)$. (Hint: Use Proposition 3.2.5.)

(ii) Prove the following formula (e.g., first for $u \in C_0^\infty(\mathbf{R}^n)$):

$$\begin{aligned} u(y) &= \frac{1}{2^{\frac{n}{2}}\pi^{\frac{3n}{4}}} \int_{\substack{\omega\in S^{n-1}\\ h>0\\ x\in\mathbf{R}^n}} e^{i(y-x)\omega/h-(x-y)^2/2h} \\ &\qquad\qquad \times Tu(x,\omega\,;\,h)h^{-1-\frac{3n}{2}}\,dx\,dh\,d\omega. \end{aligned}$$

(Hint: Integrate first with respect to x, then make a change of contour of integration of the type $S^{n-1} \ni \omega \mapsto \omega - i(y-y')/4$, and the change of variable $\lambda = h^{-1}$.)

(iii) Deduce from (ii) that if $\{x_0\} \times \mathbf{R}^n \cap \mathrm{WF}_a(u) = \emptyset$, then u is analytic near x_0.

(iv) Use the previous result to prove that if $(x_0, 0) \notin \mathrm{MS}(u)$, then $x_0 \notin \mathrm{Supp}\,u$. (Hint: First use (i) to get that $\{x_0\} \times \mathbf{R}^n \cap \mathrm{MS}(u) = \emptyset$; then deduce the analyticity of u near x_0 and use it to prove the estimate (4.1.6) by making in the expression of $Tu(x,\xi)$ the change of contour of integration $\mathbf{R}^n \ni y \mapsto y - i\delta\chi(y)\xi/|\xi|$ with $\delta > 0$ small enough and χ supported near x_0; finally, apply Corollary 4.1.3 (iii) and use the fact that u does not depend on h.)

(v) Deduce from (iv) the identity

$$\mathrm{MS}(u) = \mathrm{WF}_a(u) \cup \mathrm{Supp}\,u \times \{0\}.$$

5. Assume that $V \in C^\infty(\mathbf{R}^n\,;\,\mathbf{R})$ satisfies $V(x) \geq -C$ for all $x \in \mathbf{R}^n$ and for some constant $C > 0$. Then, setting $p(x,\xi) = \xi^2 + V(x)$, prove that $\exp tH_p(x,\xi)$ exists for all $t \in \mathbf{R}$ and for all $(x,\xi) \in \mathbf{R}^{2n}$. (Hint: Use the conservation of the total energy to show that $\exp tH_p(x,\xi)$ remains uniformly bounded on any bounded interval of time $(-T_-, T_+)$ where it

exists; deduce that $t \mapsto \exp tH_p(x,\xi)$ is uniformly Lipschitz as $t \to \pm T_\pm$, and then that $\exp tH_p(x,\xi)$ admits a limit $(x_\pm, \xi_\pm)$ as $t \to \pm T_\pm$; conclude by considering the solution of the Hamilton–Jacobi equations with initial value $(x_\pm, \xi_\pm)$.)

6. **Hanges Theorem** - The purpose of this problem is to generalize Theorem 4.3.7 to the case where p_0 is not necessarily real, assuming only that $\gamma(t) := \exp tH_{p_0}(x_0, \xi_0)$ is real for $t \in]T_0, T_1[$. As for Theorem 4.3.7, we can assume that $p \in S_{2n}^{hol}(1)$, and reasoning by contradiction we have to deduce from (4.3.7) that $(x_0, \xi_0) \notin \mathrm{MS}(u)$. For $\varepsilon > 0$ small enough, consider the solution $\psi_t = \psi_t(x, \xi; \varepsilon)$ of the system

$$\begin{cases} \varepsilon\partial_t\psi_t = -\chi(x,\xi)\mathrm{Im}p_0(x - 2\varepsilon\partial_z\psi_t, \xi + 2i\varepsilon\partial_z\psi_t), \\ \psi_t\,|_{t=0} = \varphi_\delta. \end{cases}$$

where $\chi \in C_0^\infty$ is a cutoff function supported around a fixed segment of γ containing (x_0, ξ_0), and φ_δ is supported near $(x_\delta, \xi_\delta) := \exp t_\delta H_{p_0}(x_0, \xi_0)$ ($\delta > 0$ fixed small enough) and is choosen in such a way that $\varphi_\delta(x_\delta, \xi_\delta) > 0$ and $\|e^{\varphi_\delta/h}Tu\| = \mathcal{O}(1)$.

(i) For $(x,\xi) \in \gamma$, set

$$\phi_t(x,\xi) := \psi_t(x,\xi) - \varphi_\delta(\exp -tH_{p_0}(x,\xi)).$$

By a Taylor expansion with respect to ε, prove that

$$\left(\frac{\partial}{\partial t} + H_{p_0}\right)\phi_t(x,\xi) = (1 - \chi(x,\xi))H_{p_0}\psi_t + \mathcal{O}_\delta(\varepsilon),$$

where the (δ-dependent) $\mathcal{O}_\delta(\varepsilon)$ is uniform with respect to ε small enough, $|t| \le t_0$ with t_0 independent of δ, and $(x,\xi) \in \gamma \cap W_0$ with W_0 some fixed neighborhood of (x_0, ξ_0).

(ii) Using that $\phi_t\,|_{t=0} = 0$, deduce from the previous question that for $(x,\xi) \in \{\chi = 1\} \cap \gamma \cap W_0$ one has

$$\phi_t = \mathcal{O}_\delta(\varepsilon)$$

uniformly with respect to (x,ξ), $|t| \le t_0$ and ε small enough. In particular, prove that if ε is chosen sufficiently small with respect to δ, then $\psi_t(\exp(tH_{p_0}(x_\delta, \xi_\delta)) > 0$ for all $|t| \le t_0$.

(iii) Setting

$$f(t) = \|e^{\varepsilon\psi_t/h}Tu\|^2$$

and applying corollary 3.5.3, show that

$$\begin{aligned} hf'(t) &= -2\mathrm{Im}\,\langle \chi e^{\varepsilon\psi_t/h}TPu, e^{\varepsilon\psi_t/h}Tu\rangle + \mathcal{O}(h)\|e^{\varepsilon\psi_t/h}Tu\|^2 \\ &= \mathcal{O}(h)f(t). \end{aligned}$$

(iv) Deduce that there exists a positive constant C independent of δ such that

$$f(t) \leq Ce^{C|t|}f(0)$$

for $|t| \leq C^{-1}$, and then conclude by taking $t = -t_\delta$ and by using (ii).

7. **Propagation of the Frequency Set** - The purpose of this exercise is to prove a result of propagation for FS(u) analogous to Theorem 4.3.7 in the case where $u \in L^2(\mathbf{R}^n)$ is solution of

$$\begin{cases} Pu = 0, \\ \|u\| \leq 1, \end{cases}$$

with $P = \mathrm{Op}_h^t(p)$, $p = p_0 + hr$, $p_0, r \in S_{2n}(\langle\xi\rangle^m)$, p_0 real-valued independent of h. As for Theorem 4.3.7 one can assume without loss of generality that $m = 0$, and by the same arguments leading to (4.3.7) one has to prove that if for all $\delta > 0$ there exists $t \in [-\delta, \delta]$ such that $Tu = \mathcal{O}(h^\infty)$ in a neighborhood W_δ of $(x_\delta, \xi_\delta) := \exp t_\delta H_{p_0}(x_0, \xi_0)$, then $(x_0, \xi_0) \notin$ FS(u) (see Exercise 3 of Chapter 3 for a definition of FS(u) involving Tu).

(i) Using Exercise 4 of Chapter 3 and (4.3.9), prove that for any $\psi \in C_0^\infty(\mathbf{R}^{2n})$ real-valued one has

$$\mathrm{Im}\left\langle h^\psi TPu, h^\psi Tu\right\rangle = h\ln h\left\langle (H_{p_0}\psi)h^\psi Tu, h^\psi Tu\right\rangle + \mathcal{O}(h)\|h^\psi Tu\|^2$$

uniformly for $h > 0$ small enough.

(ii) Let g be the function constructed in (4.3.13) with $\alpha_\delta = 1$. Taking $\psi := -Ng$ with $N > 0$ fixed arbitrarily large and using that $Pu = 0$, deduce from (i) that (with the same notation as in (4.3.13)) one has

$$\|h^\psi Tu\|_{V_\delta}^2 = \mathcal{O}(h^{(2\beta d-1)N/2}\|Tu\|^2 + h^{-N}\|Tu\|_{W_\delta}^2 + |\ln h|^{-1}\|h^\psi Tu\|_{V_\delta}^2).$$

(iii) In particular, denoting by W a (δ-dependent) neighborhood of (x_0, ξ_0) on which $g \geq (1-\beta d)/2$, deduce from (iii) that for h small enough one has
$$h^{-\beta dN}\|Tu\|_W = \mathcal{O}(1)$$
uniformly with respect to h.

(iv) Conclude that $(x_0, \xi_0) \notin \mathrm{FS}(u)$ and state the result of propagation for $\mathrm{FS}(u)$.

8. Prove the result of Remark 4.3.14. (Hint: First prove that if $\chi = 1$ near $\Pi_x \mathrm{Supp}\, g$ and θ is small enough, then $\left\| e^{\theta g/h} T \langle hD_x \rangle^s [P, \chi] u \right\| = \mathcal{O}(1)$; then observe that $v := \langle hD_x \rangle^s \chi u$ is solution of $Qv = \langle hD_x \rangle^s [P, \chi] u$ with $Q := P + [\langle hD_x \rangle^s, P] \langle hD_x \rangle^{-s} = P + \mathcal{O}(h)$.)

9. **Propagation in Evolution Problems** - Let $u \in C^1(\mathbf{R}_t\, ;\, L^2(\mathbf{R}^n_x))$ be a solution of an evolution equation of the type
$$(hD_t + P)u = 0,$$
where $P = \mathrm{Op}_h^W(p)$ is a semiclassical pseudodifferential operator with symbol $p = p_0 + \varepsilon(h) r \in S_{2n}^{hol}(1)$, $\varepsilon(h) \to 0$ as $\varepsilon \to 0$, $p_0 \,|_{\mathbf{R}^{2n}}$ real-valued. The purpose of this problem is to show that for all $(t_0, x_0, \xi_0) \in \mathbf{R} \times \mathbf{R}^{2n}$ one has the following equivalence:
$$(x_0, \xi_0) \in \mathrm{MS}(u\,|_{t=0}) \quad \Leftrightarrow \quad \exp t_0 H_{p_0}(x_0, \xi_0) \in \mathrm{MS}(u\,|_{t=t_0}). \qquad (4.6.1)$$
We denote by T the usual global FBI transform (with respect to the variable x), so that Tu is a function of (t, x, ξ).

(i) Assume that for all $\tau \in \mathbf{R}$, one has $(0, x; \tau, \xi) \notin \mathrm{MS}(u)$. Observing that Tu is solution of an evolution equation of the type
$$(hD_t + \widetilde{P}(x, hD_\xi, hD_x))Tu = 0$$
and using the classical ellipticity with respect to τ of the symbol $\tau + \widetilde{p}$ of $hD_t + \widetilde{P}$, prove that $Tu(t, x, \xi)$ is uniformly exponentially small in a neighborhood of $(0, x_0, \xi_0)$. In particular, deduce that $(x_0, \xi_0) \notin \mathrm{MS}(u\,|_{t=0})$.

(ii) Conversely, assume that $(x_0, \xi_0) \notin \mathrm{MS}(u\,|_{t=0})$. Then considering $w := e^{-it\widetilde{P}/h} Tu$ (where $e^{-it\widetilde{P}/h}$ is defined by its convergent Taylor series), prove that w does not depend on t and deduce that Tu is uniformly exponentially small in a neighborhood of $(0, x_0, \xi_0)$.

(iii) Deduce from (i) and (ii) that the following equivalence holds:

$$(x_0, \xi_0) \in \mathrm{MS}(u\,|_{t=0}) \;\Leftrightarrow\; \exists\, \tau_0 \in \mathbf{R} \text{ with } (0, x_0, \tau_0, \xi_0) \in \mathrm{MS}(u).$$

(iv) Applying the local version of Theorem 4.3.7 given in Remark 4.3.14 (see also Exercise 8 of this chapter) to the operator $Q := hD_t + P$ (which is differential with respect to t), prove that if $(0, x_0, \tau_0, \xi_0) \in \mathrm{MS}(u)$, then for all $t \in \mathbf{R}$ one has $(t, x_t, \tau_0, \xi_t) \in \mathrm{MS}(u)$, where $(x_t, \xi_t) := \exp tH_{p_0}(x_0, \xi_0)$.

(v) Applying (iii) at $t = t_0$ instead of $t = 0$ (and explaining why this can be done), deduce from (iv) that (4.6.1) holds.

(vi) Application: Let $\Phi^t_{x,\xi} = e^{-itP/h}\Phi_{x,\xi}$ be the quantum evolution of the coherent state $\Phi_{x,\xi}$ centered at (x, ξ) (see Exercise 1 of Chapter 3). Prove that $\mathrm{MS}(\Phi^t_{x,\xi}) = \{\exp tH_p(x, \xi)\}$. Generalize this result to $p \in S^{hol}_{2n}(\langle\xi\rangle^m)$ with $m > 0$ under the additional assumption that the whole symbol $p\,|_{\mathbf{R}^{2n}}$ is real-valued.

Note: Observe that Theorem 4.3.7 can be seen as a particular case of (4.6.1) corresponding to u independent of t.

10. **Exact Egorov Theorem for the Harmonic Oscillator** - Let $H_0 = -h^2\Delta + \sum_{j=1}^{n} \omega_j^2 x_j^2$ (with $\omega_j \geq 0$). Prove that for any $t \in \mathbf{R}$ and $p \in S_{2n(1)}$ one has

$$e^{itH_0/h}\mathrm{Op}_h^W(p)e^{-itH_0/h} = \mathrm{Op}_h^W(p \circ \phi_t),$$

where ϕ_t is the Hamilton flow associated with H_0. (Hint: Observe that both operators are solutions of the same evolution problem.)

11. **Egorov Theorem** - Find a generalization of the previous result when H_0 is replaced by any symmetric h-pseudodifferential operator H. (Hint: Set $R(t) := e^{-itH/h}\mathrm{Op}_h^W(p \circ \phi_t)e^{itH/h} - \mathrm{Op}_h^W(p)$, compute $\partial_t R(t)$, and deduce that $\|R(t)\|$ is $\mathcal{O}(h)$. Then iterate the procedure. The final result is that $e^{itH/h}\mathrm{Op}_h^W(p)e^{-itH/h} = \mathrm{Op}_h^W(\widetilde{p}_t) + \mathcal{O}(h^\infty)$ in $L^2(\mathbf{R}^n)$, with $\widetilde{p}_t \sim p \circ \phi_t + \sum_{j\geq 1} h^j p_{j,t}$ in $S_{2n}(1)$.)

12. **Propagation of the Frequency Set by Means of Egorov Theorem** - Using Definition 2.9.1 and applying the previous exercise with $p = \chi \in C_0^\infty(\mathbf{R}^{2n})$, prove that for all $u \in L^2(\mathbf{R}^n)$ one has $\mathrm{FS}\left(e^{itH/h}u\right) =$

$\phi_t(\mathrm{FS}(u))$, where ϕ_t denotes the Hamilton flow associated with the self-adjoint h-pseudodifferential operator H.

Chapter 5

Complements: Symplectic Aspects

5.1 The Canonical Symplectic Form on $\mathbf{R}^{2n}$

For $u = (u_x, u_\xi)$ and $v = (v_x, v_\xi) \in \mathbf{R}^{2n}$, we set

$$\sigma(u,v) = u_\xi \cdot v_x - u_x \cdot v_\xi = \left\langle \begin{pmatrix} 0 & I \\ -I & 0 \end{pmatrix} \begin{pmatrix} u_x \\ u_\xi \end{pmatrix}, \begin{pmatrix} v_x \\ v_\xi \end{pmatrix} \right\rangle.$$

This formula defines a bilinear map: $\mathbf{R}^{2n} \times \mathbf{R}^{2n} \to \mathbf{R}$, which moreover is antisymmetric, in that

$$\sigma(u,v) = -\sigma(v,u) \quad \text{for all } u, v \in \mathbf{R}^{2n},$$

and nondegenerate:

$$\text{If } \sigma(u,v) = 0 \text{ for all } v \in \mathbf{R}^{2n}, \quad \text{then } u = 0.$$

The map σ is called the *canonical symplectic 2-form* (or more simply the *canonical symplectic form*) *on* $\mathbf{R}^{2n}$, and it is also denoted by

$$\sigma = \sum_{j=1}^{n} d\xi_j \wedge dx_j = d\xi \wedge dx.$$

In particular, we have

$$\sigma = d\omega, \tag{5.1.1}$$

where d denotes the operator of differentiation, and $\omega \in C^\infty(\mathbf{R}^{2n}; \mathcal{L}(\mathbf{R}^{2n}; \mathbf{R}))$ is the so-called *canonical 1-form on* $\mathbf{R}^{2n}$, defined as follows:

$$\text{For } (x,\xi) \text{ and } (u_x, u_\xi) \in \mathbf{R}^{2n}, \quad \omega_{(x,\xi)}(u_x, u_\xi) = \sum_{j=1}^{n} \xi_j u_{x_j}.$$

More synthetically, ω is also written

$$\omega = \sum_{j=1}^{n} \xi_j dx_j,$$

and the identity (5.1.1) shows that σ is *exact* in the sense of the 2-forms.

Actually, σ must be interpreted as an element of $C^\infty(\mathbf{R}^{2n}_\rho; \mathcal{B}_2(\mathbf{R}^{2n}; \mathbf{R}))$ (where $\mathcal{B}_2(\mathbf{R}^{2n}; \mathbf{R})$ denotes the space of bilinear forms on $\mathbf{R}^{2n}$), which appears to be constant with respect to $\rho \in \mathbf{R}^{2n}$:

$$\forall \rho \in \mathbf{R}^{2n}, \quad \sigma_\rho = d\xi \wedge dx.$$

Then the fact that σ is exact implies, in particular, that it is also *closed*, that is,

$$d\sigma = 0.$$

Of course, in such a framework the space $\mathbf{R}^{2n}$ appearing in the expression $\mathcal{B}_2(\mathbf{R}^{2n}; \mathbf{R})$ has also to be interpreted in a special way, namely, as the *tangent space* $T_\rho(\mathbf{R}^{2n})$. Indeed, in all our considerations, the space $\mathbf{R}^{2n}_\rho$ could be replaced by any $2n$-dimensional manifold M, in which case $\mathcal{B}_2(\mathbf{R}^{2n}; \mathbf{R})$ has to be replaced by $\mathcal{B}_2(T_\rho M; \mathbf{R})$, where $T_\rho M$ is the vector space tangent to M at ρ.

However, for the sake of simplicity (and since we have not considered more complicated situations in these notes), we shall continue to work in the vector space $\mathbf{R}^{2n}$. The reader interested in more general considerations may consult, e.g., [St].

5.2 Symplectic Transformations

Let $\mathcal{U}$ and $\mathcal{V}$ be two open subsets of $\mathbf{R}^{2n}$, and let

$$\begin{array}{rccc} \kappa : & \mathcal{U} & \to & \mathcal{V}, \\ & (x,\xi) & \mapsto & (y(x,\xi), \eta(x,\xi)), \end{array}$$

be a C^∞ diffeomorphism from $\mathcal{U}$ to $\mathcal{V}$. One says that κ is *canonical* (or *symplectic*) if one has on $\mathcal{U}$

$$d\eta \wedge dy = d\xi \wedge dx,$$

where we have set

$$d\eta\wedge dy = \sum_{j=1}^{n} d\eta_j\wedge dy_j = \sum_{j=1}^{n}\left(\sum_{i=1}^{n}\frac{\partial\eta_j}{\partial x_i}dx_i + \frac{\partial\eta_j}{\partial\xi_i}d\xi_i\right)\wedge\left(\sum_{k=1}^{n}\frac{\partial y_j}{\partial x_k}dx_k + \frac{\partial y_j}{\partial\xi_k}d\xi_k\right).$$

(On the right-hand side of this last formula, $\wedge$ stands for the usual exterior product of 1-forms on $\mathbf{R}^{2n}$.)

This is also equivalent to

$$\boxed{\kappa^*\sigma = \sigma,} \tag{5.2.1}$$

where $(\kappa^*\sigma)_\rho(u,v) := \sigma(d\kappa(\rho)\cdot u, d\kappa(\rho)\cdot v)$.

In other words, κ preserves σ. In the particular case where $n = 1$, $-\sigma$ is just the determinant in $\mathbf{R}^2$ (that is, the oriented length of the vectorial product), and therefore in this case (5.2.1) means that κ preserves the areas of $\mathbf{R}^2$.

Example - Let

$$\begin{array}{ccc}\mathbf{R}^n & \to & \mathbf{R}^n,\\ x & \mapsto & y(x),\end{array}$$

be a change of variables, and set:

$$\kappa(x,\xi) = (y(x), \eta(x,\xi))$$

with

$$\eta(x,\xi) = {}^t dy(x)^{-1}\xi.$$

Then, setting $M = M(x) = {}^t dy(x)^{-1}$, one has

$$d\eta = M(x)d\xi + \frac{\partial M\xi}{\partial x}dx,$$

and thus, with $N = {}^t dy(x)* = M^{-1}$,

$$\begin{aligned}dy\wedge d\eta &= {}^tN(x)dx\wedge M(x)d\xi + {}^tN(x)dx\wedge\frac{\partial M\xi}{\partial x}dx\\ &= dx\wedge N(x)M(x)d\xi + dx\wedge N(x)\frac{\partial M\xi}{\partial x}dx\\ &= dx\wedge d\xi + \sum_{j=1}^{n} dx_j\wedge\left(\sum_{k=1}^{n} a_{j,k}(x,\xi)dx_k\right)\\ &= dx\wedge d\xi + \sum_{j<k}(a_{j,k}(x,\xi) - a_{k,j}(x,\xi))dx_j\wedge dx_k,\end{aligned} \tag{5.2.2}$$

where $a_{j,k}(x,\xi)$ is the coefficient (j,k) of the matrix $N(x)\frac{\partial M\xi}{\partial x}$. Since this matrix depends linearly on ξ, a way of proving that it is symmetric is to verify it when ξ is any vector of the canonical basis of $\mathbf{R}^n$. Assume, e.g., that $\xi = (1,0,\dots,0)$. In this case, denoting by $m_{i,j}$ and $n_{i,j}$ the matrix coefficients of M and N, respectively, one has

$$a_{j,k} = \sum_{\ell=1}^{n} n_{j,\ell}\frac{\partial m_{\ell,1}}{\partial x_k},$$

and thus by differentiation of the identity $\sum_{\ell=1}^{n} n_{j,\ell}m_{\ell,1} = \delta_{j,1}$ (which is a consequence of the fact that $MN = I$), one gets

$$a_{j,k} = -\sum_{\ell=1}^{n}\frac{\partial n_{j,\ell}}{\partial x_k}\partial m_{\ell,1} = -\sum_{\ell=1}^{n}\frac{\partial^2 y_\ell}{\partial x_k \partial x_j}\partial m_{\ell,1},$$

where the last step is a direct consequence of the definition of N. This expression of $a_{j,k}$ obviously remains invariant in exchanging j and k, and since the same computation works for any vector ξ of the canonical basis of $\mathbf{R}^n$, this proves that $a_{j,k}(x,\xi) = a_{k,j}(x,\xi)$ on $\mathbf{R}^{2n}$. As a consequence, going back to (5.2.2), one gets

$$dy \wedge d\eta = dx \wedge d\xi,$$

and therefore κ is canonical.

Other examples are given by the transformations j_B, k_C and ℓ_j introduced in Section 3.4.

5.3 Hamilton Field

Let $f = f(x,\xi) \in C^\infty(\mathbf{R}^{2n})$. Then for all $(x,\xi) \in \mathbf{R}^{2n}$, the differential $df(x,\xi)$ of f at (x,ξ) belongs to $\mathcal{L}(\mathbf{R}^{2n};\mathbf{R}) = (\mathbf{R}^{2n})^* \simeq T^*_{x,\xi}\mathbf{R}^{2n}$; that is, the application

$$\mathbf{R}^{2n} \simeq T_{(x,\xi)}\mathbf{R}^{2n} \ni u \mapsto \langle df(x,\xi), u\rangle$$

is a linear form on $T_{(x,\xi)}\mathbf{R}^{2n}$. Therefore, since the 2-form σ is nondegenerate, the Riesz theorem gives the existence and the unicity of $H_f(x,\xi) \in T_{x,\xi)}\mathbf{R}^{2n}$ such that for all $u \in T_{(x,\xi)}\mathbf{R}^{2n}$,

$$\boxed{\langle df(x,\xi), u\rangle = \sigma(u, H_f(x,\xi)).}$$

In Cartesian coordinates, one verifies easily that

$$H_f = \frac{\partial f}{\partial \xi}\frac{\partial}{\partial x} - \frac{\partial f}{\partial x}\frac{\partial}{\partial \xi},$$

which corresponds to the definition of the Hamilton field given in Chapter 4. The previous considerations show that this notion is indeed a *symplectic notion*; that is, it is not related to the particular choice of coordinates, but only to the symplectic form σ. In particular, H_f will have the same expression in any system of *symplectic coordinates*, that is, coordinates (y, η) such that $\sigma = d\eta \wedge dy$.

Remark 5.3.1 The Poisson bracket of two functions f and g can also be written

$$\{f, g\} = \sigma(H_f, H_g).$$

As a consequence, it is a symplectic notion, too. In particular, if κ is a symplectic transformation, then $\{f \circ \kappa, g \circ \kappa\} = \{f, g\} \circ \kappa$ for all f, g.

An important feature of the Hamilton flow is the following result:

Proposition 5.3.2 *Let $f \in C^\infty(\mathbf{R}^{2n})$ be real-valued, $\mathcal{U}$ a bounded open set of $\mathbf{R}^{2n}$, and let $T_\pm > 0$ be such that $\exp tH_f(x, \xi)$ is well-defined for $t \in (-T_-, T_+)$ and $(x, \xi) \in \mathcal{U}$. Then for all $t \in (-T_-, T_+)$, the application*

$$\mathcal{U} \ni (x, \xi) \mapsto \exp tH_f(x, \xi) \in \exp tH_f(\mathcal{U})$$

is a canonical transformation.

Proof Set $\phi_t(x, \xi) = \exp tH_f(x, \xi)$. Then one has

$$\begin{aligned}\frac{\partial}{\partial t}(\phi_t^*\sigma(u, v)) &= \frac{\partial}{\partial t}\sigma(d\phi_t \cdot u, d\phi_t \cdot v)\\ &= \sigma(dH_f(\phi_t) \cdot u, d\phi_t \cdot v) + \sigma(d\phi_t \cdot u, dH_f(\phi_t) \cdot v).\end{aligned} \tag{5.3.1}$$

But for any $\lambda \in \mathbf{R}^{2n}$, one has

$$\sigma(\lambda, H_f) = \langle df, \lambda \rangle,$$

and therefore, applying the exterior differential operator $d = d_{(x,\xi)}$ to this equality, one gets for all $\lambda, \mu \in \mathbf{R}^{2n}$ that

$$d[\sigma(\lambda, H_f)] \cdot \mu = \left\langle d^2 f \cdot \mu, \lambda \right\rangle;$$

that is, since $d^2f = 0$,

$$(d\sigma)(\lambda, H_f, \mu) + \sigma(\lambda, dH_f \cdot \mu) = 0,$$

and thus, since $d\sigma = d^2\omega = 0$,

$$\sigma(\lambda, dH_f \cdot \mu) = 0.$$

One also has $\sigma(dH_f \cdot \lambda, \mu) = -\sigma(\mu, dH_f \cdot \lambda) = 0$ for all $\lambda, \mu \in \mathbf{R}^{2n}$, and in particular, we obtain from (5.3.1),

$$\frac{\partial}{\partial t}(\phi_t^* \sigma) = 0.$$

As a consequence, $\phi_t^*\sigma = \phi_t^*\sigma \,|_{t=0} = \sigma$. ◇

5.4 Lagrangian Submanifolds

Definition 5.4.1 *A submanifold Λ of $\mathbf{R}^{2n}$ is said to be* **Lagrangian** *if its dimension is n and if $\sigma\,|_\Lambda = 0$ (that is, for any $\rho \in \Lambda$, $\sigma\,\Big|_{T_\rho\Lambda} = 0$).*
Equivalently, Λ is Lagrangian if for all $\rho \in \Lambda$, one has

$$(T_\rho\Lambda)^{\perp_\sigma} = T_\rho\Lambda,$$

where $(T_\rho\Lambda)^{\perp_\sigma}$ denotes the orthogonal space of $T_\rho\Lambda$ in $\mathbf{R}^{2n}$ with respect to the 2-form σ.

Remark 5.4.2 The previous equivalence is a direct consequence of the non-degeneracy of σ on $\mathbf{R}^{2n}$.

Remark 5.4.3 One deduces immediately from this definition and (5.2.1) that if κ is a canonical transformation defined on a neighborhood of a Lagrangian submanifold Λ, then $\kappa(\Lambda)$ is Lagrangian, too.

Example - For any arbitrary C^∞ function $\varphi : \mathbf{R}^n \to \mathbf{R}$, set

$$\Lambda_\varphi := \{(x, \xi) \in \mathbf{R}^{2n} \; ; \; \xi = \nabla\varphi(x)\}.$$

Then Λ_φ is a Lagrangian submanifold of $\mathbf{R}^{2n}$, since we have

$$\begin{aligned} dx \wedge d\xi\Big|_{\Lambda_\varphi} &= \sum_i dx_i \wedge d\xi_i\Big|_{\Lambda_\varphi} = \sum_i \left(dx_i \wedge \sum_j \frac{\partial^2 \varphi}{\partial x_j \partial x_i} dx_j \right) \\ &= \sum_{i<j} \frac{\partial^2 \varphi}{\partial x_j \partial x_i} (dx_i \wedge dx_j + dx_j \wedge dx_i) = 0. \end{aligned}$$

Of course, the same is true for $\tilde{\Lambda}_\varphi := \{(x,\xi)\ ;\ x = \nabla\varphi(\xi)\}$.

Actually, these two examples have somehow a general character because of the following important result:

Proposition 5.4.4 *Let Λ be a submanifold of $\mathbf{R}^{2n}$ of dimension n such that the projection*

$$\begin{array}{rccc} \pi_1 : & \Lambda & \to & \mathbf{R}^n, \\ & (x,\xi) & \mapsto & x, \end{array}$$

is a local diffeomorphism. Then Λ is Lagrangian if and only if Λ can be locally described by an equation of the type $\xi = \nabla\varphi(x)$ with $\varphi : \mathbf{R}^n \to \mathbf{R}$ smooth.

Proof By assumption we already know that Λ is locally defined by an equation of the type $\xi = F(x)$ for some smooth vectorial function F. Now, saying that Λ is Lagrangian is equivalent to saying that $\sigma|_\Lambda = d\omega|_\Lambda = 0$; that is, the 1-form $\omega|_\Lambda$ is closed. This, in turns is equivalent to saying that $\xi dx|_\Lambda$ is locally exact; that is, $F(x)dx = d\varphi$ for some locally defined smooth function $\varphi = \varphi(x)$. This proves the result. ◇

5.5 Applications to the Study of PDEs

(a) Solutions of Eikonal Equations

Let $p \in C^\infty(\mathbf{R}^{2n})$. If one tries to solve formally and locally the equation

$$\mathrm{Op}_h^t(p)\left(a(x;h)e^{i\varphi(x)/h}\right) = 0$$

with $a \sim \sum_{j\geq 0} h^j a_j(x)$ and $\varphi \in C^\infty$, the first equation one gets is the so-called *eikonal equation:*

$$p(x, \nabla\varphi(x)) = 0. \tag{5.5.1}$$

The general idea for solving this equation is not to try to construct φ directly, but rather to construct in a geometric way the Lagrangian manifold $\Lambda_\varphi = \{\xi = \nabla\varphi(x)\}$ associated with φ. Indeed, equation (5.5.1) is obviously equivalent to

$$\Lambda_\varphi \subset p^{-1}(0). \tag{5.5.2}$$

Then the first observation one can make is that (5.5.2) implies that for any $\rho \in \Lambda_\varphi$, one has

$$(T_\rho p^{-1}(0))^{\perp_\sigma} \subset (T_\rho \Lambda_\varphi)^{\perp_\sigma} = T_\rho \Lambda_\varphi,$$

and since $(T_\rho p^{-1}(0))^{\perp_\sigma}$ is one-dimensional and generated by $H_p(\rho)$ (because $\sigma(u, H_p(\rho)) = \langle dp(\rho), u\rangle$), the vector $H_p(\rho)$ must necessarily be tangent to Λ_φ. In other words, the vector field H_p remains tangent to any solution Λ_φ of (5.5.2). The following result roughly says that this condition is also essentially sufficient, and it actually gives a way to construct many solutions of (5.5.2).

Theorem 5.5.1 *Let $\Lambda' \subset p^{-1}(0)$ be a submanifold of dimension $n-1$ such that $\sigma|_{\Lambda'} = 0$ and H_p is transverse to Λ'. Then for any $\rho'_0 \in \Lambda'$, there exist $t_0 > 0$ and a neighborhood Λ'_0 of ρ in Λ' such that the set*

$$\Lambda = \bigcup_{|t|<t_0} \exp tH_p(\Lambda'_0)$$

is a Lagrangian submanifold of $\mathbf{R}^{2n}$ that satisfies

$$\Lambda' \subset \Lambda \subset p^{-1}(0).$$

Proof The transversality of H_p with respect to Λ' gives that Λ is a submanifold of $\mathbf{R}^{2n}$ of dimension n if Λ'_0 and t_0 are taken small enough (a possible parametrization of Λ is then given by $\{(t, y(\rho'))\ ;\ |t| < t_0\ ,\ \rho' \in \Lambda'_0\}$, where $\rho' \mapsto y(\rho') \in \mathbf{R}^{n-1}$ is a parametrization of Λ' near ρ'_0). Moreover, if $\rho = \exp tH_p(\rho') = \phi_t(\rho') \in \Lambda$, one has

$$\sigma|_{T_\rho\Lambda} = \phi_t^*\sigma|_{T_{\rho'}\Lambda' \oplus \langle H_p(\rho')\rangle},$$

where $\langle H_p(\rho')\rangle$ denotes the one-dimensional vector space generated by $H_p(\rho')$. Then applying Proposition 5.3.2, we get

$$\sigma|_{T_\rho\Lambda} = \sigma|_{T_{\rho'}\Lambda' \oplus \langle H_p(\rho')\rangle}. \tag{5.5.3}$$

Now if $u, v \in T_{\rho'}\Lambda'$, we have $\sigma(u,v) = 0$ by assumption on Λ', and also $\sigma(u, H_p(\rho')) = \langle dp(\rho'), u\rangle = 0$ (and the same for v), since $\Lambda' \subset p^{-1}(0)$. Since, moreover, $\sigma(H_p(\rho'), H_p(\rho')) = 0$, we finally get

$$\sigma(u + H_p(\rho'), v + H_p(\rho')) = 0,$$

which leads to the result by (5.5.3). ◇

Remark 5.5.2 Any submanifold $\Lambda' \subset \mathbf{R}^{2n}$ such that $\sigma|_{\Lambda'} = 0$ (or equivalently, $T_\rho\Lambda' \subset (T_\rho\Lambda')^{\perp_\sigma}$ for all $\rho \in \Lambda'$) is said to be *isotropic*. In particular, its dimension is then at most n. If on the contrary $(T_\rho\Lambda')^{\perp_\sigma} \subset T_\rho\Lambda'$, then Λ' is said to be *involutive*, and its dimension is at least n.

Coming back to our initial problem, an interesting consequence of the previous result is the following:

Corollary 5.5.3 *If $p(x_0, \xi_0) = 0$, $\dfrac{\partial p}{\partial \xi_n}(x_0, \xi_0) \neq 0$, and $\psi = \psi(x')$ is C^∞ near x_0' such that*

$$\frac{\partial \psi}{\partial x'}(x_0') = \xi_0',$$

then there exists a function $\varphi = \varphi(x)$ smooth near x_0 that satisfies

$$\varphi'(x_0) = \xi_0;\ \varphi(x', 0) = \psi(x');\ p(x, \nabla\varphi_x) = 0 \text{ near } x_0.$$

Proof Just apply the previous theorem with

$$\Lambda' = \left\{(x, \xi)\ ;\ x_n = x_n^0\ ,\ \xi' = \frac{\partial \psi}{\partial x'}(x')\ ,\ \xi_n = \lambda(x', \xi')\right\},$$

where the function λ is defined as the local implicit solution of the problem

$$p(x', x_n^0; \xi) = 0 \Leftrightarrow \xi_n = \lambda(x', \xi').$$

◇

(b) Canonical Transformations of Pseudodifferential Operators

Let $p = p(x, \xi) \in S(1)$ be the symbol of a pseudodifferential operator you want to study. In general, p can look very complicated, but assume that there exists a symplectic transformation κ such that the function $p \circ \kappa$ is much simpler near some point (y_0, η_0) of $\mathbf{R}^{2n}$ (a typical example is when $\nabla p(x_0, \xi_0) \neq 0$; in this case, one can construct κ symplectic such that $p \circ \kappa(y, \eta) = \eta_n$ near $(y_0, \eta_0) := \kappa^{-1}(x_0, \xi_0)$). Since one wants to study not the function p, but a quantization P of p, a natural question is then to know whether the existence of such a κ makes it possible also to simplifythe operator P (e.g., by transforming it into an operator of symbol $p \circ \kappa$).

The answer we are going to give is essentially positive, in the sense that there exists an operator A such that we formally have

$$APA^{-1} = \mathrm{Op}_h(p \circ \kappa) + \mathcal{O}(h)$$

microlocally near (y_0, η_0), in a sense that will be specified.

Another related problem is just the opposite one: Assume that you have an operator A of some special type (typically, a Fourier integral operator) and that is somehow invertible. Then what can you say about APA^{-1} if P is a pseudodifferential operator? In particular, is it a pseudodifferential operator, too, and with which symbol? This problem will get a positive answer, too, and all the results concerning such situations are usually called *Egorov theorems* from the original paper [Eg].

Let us start by giving some examples where all this can be done in a simple way.

Example 1 - Let $\kappa : (y, \eta) \mapsto (y, \eta - \eta_0)$, where η_0 is fixed in $\mathbf{R}^n$, and define the operator A by

$$(Au)(y) = e^{iy\eta_0/h} u(y).$$

Then A is invertible (e.g., on $L^2(\mathbf{R}^n)$), and it is easy to see that if $P = \mathrm{Op}_h^t(p)$ with $t \in [0, 1]$ arbitrary, then the operator $\widetilde{P} := APA^{-1}$ satisfies

$$\widetilde{P} = e^{iy\eta_0/h}\mathrm{Op}_h^t(p)e^{-iy\eta_0/h} = \mathrm{Op}_h^t(p(y, \eta - \eta_0)) + \mathcal{O}(h) = \mathrm{Op}_h^t(p \circ \kappa(y, \eta)) + \mathcal{O}(h).$$

Example 2 - Let $f : y \mapsto f(y)$ be a global change of variables on $\mathbf{R}^n$, with, e.g., $\partial^\alpha f(y) = \mathcal{O}(1)$ uniformly on $\mathbf{R}^n$ for all $\alpha \neq 0$. Then the transformation

$$\kappa : (y, \eta) \mapsto (f(y), {}^t df(y)^{-1}\eta) \tag{5.5.4}$$

is symplectic on $\mathbf{R}^{2n}$, and a direct computation together with the results of Chapter 2 show that the operator

$$Au(x) := u \circ f(x)$$

satisfies

$$A\mathrm{Op}_h^t(p)A^{-1} = \mathrm{Op}_h^t(p \circ \kappa) + \mathcal{O}(h)$$

for any $p \in S_{2n}(1)$. In the sequel, the transformation κ defined in (5.5.4) will be referred to as the canonical transformation associated with the change of variables f.

Example 3 - For $t \in \mathbf{R}$, denote by $A_t = e^{itH_0/h}$ the quantum flow of $H_0 = -h^2\Delta + \sum_{j=1}^{n} \omega_j^2 x_j^2$ $(\omega_j \geq 0)$. Then one can prove easily (see Exercise 10 of Chapter 4) that for any $p \in S_{2n}(1)$ one has

$$A_t \mathrm{Op}_h^W(p) A_t^{-1} = \mathrm{Op}_h^W(p \circ \phi_t), \tag{5.5.5}$$

where ϕ_t is the Hamilton flow associated with H_0.

About the first two examples above, an observation has to be made: In both cases there exists a function $\varphi = \varphi(x, \eta)$ such that for all $(x, \eta) \in \mathbf{R}^{2n}$ one has

$$\kappa\left(\frac{\partial\varphi}{\partial\eta}(x,\eta),\eta\right) = \left(x, \frac{\partial\varphi}{\partial x}(x,\eta)\right). \tag{5.5.6}$$

Indeed, one can take $\varphi(x, \eta) = x \cdot (\eta - \eta_0)$ in the first example, and $\varphi(x, \eta) = f^{-1}(x) \cdot \eta$ in the second one. Such canonical transformations play a special role in the kind of result we are looking for, and for this reason a particular terminology is introduced for them:

Definition 5.5.4 *A canonical transformation κ is said to admit a* **generating function** *near some point $(y_0, \eta_0) = \kappa^{-1}(x_0, \xi_0)$ if there exists a smooth function φ defined in a neighborhood of (x_0, η_0) such that for all (x, η) close enough to (x_0, η_0) the identity (5.5.6) is valid and $\partial_\eta\varphi(x_0, \eta_0) = y_0$ (so that also $\partial_x\varphi(x_0, \eta_0) = \xi_0$). The function φ is called a generating function of κ.*

The first result we give concerns this type of transformation only. In the sequel, we say that two symbols of $S_{2n}(1)$ are equivalent *near* some point if their difference together with all its derivatives is $\mathcal{O}(h^\infty)$ near this point.

Theorem 5.5.5 (Egorov 1) *Let κ be a canonical transformation admitting a generating function φ near some point $(y_0, \eta_0) = \kappa^{-1}(x_0, \xi_0) \in \mathbf{R}^{2n}$. Let also $a = a(x, y, \eta; h) \in S_{3n}(1)$ be a symbol supported in a small enough neighborhood of (x_0, y_0, η_0) and elliptic at (x_0, y_0, η_0), and denote by A the operator defined on $\mathcal{D}'(\mathbf{R}^n_x)$ by*

$$Au(y\ ;\ h) = \frac{1}{(2\pi h)^n} \int e^{i[y\cdot\eta - \varphi(x,\eta)]/h} a(x, y, \eta) u(x) dx\, d\eta \in C_0^\infty(\mathbf{R}^n). \tag{5.5.7}$$

Then there exists $b = b(x, y, \eta; h) \in S_{3n}(1)$ such that if one denotes by B the operator on $\mathcal{D}'(\mathbf{R}^n_y)$ defined by

$$Bu(x\ ;\ h) = \frac{1}{(2\pi h)^n} \int e^{i[\varphi(x,\eta) - y\cdot\eta]/h} b(x, y, \eta) u(y) dy\, d\eta \tag{5.5.8}$$

then one has:

(i) For any $p \in S_{2n}(1)$ and for any $t \in [0, 1]$, the operators $A\mathrm{Op}_h^t(p)B$ and $B\mathrm{Op}_h^t(p)A$ are h-pseudodifferential operators.

(ii) For any $t \in [0, 1]$, the t-symbol $\sigma_t(AB)$ of AB satisfies

$$\sigma_t(AB) \sim 1 \quad \text{near} \quad (y_0, \eta_0).$$

(iii) For any $t \in [0, 1]$,

$$\sigma_t(BA) \sim 1 \quad \text{near} \quad (x_0, \xi_0).$$

(iv) For any $p \in S_{2n}(1)$ and for any $t \in [0, 1]$, one has

$$\sigma_t\left(A\mathrm{Op}_h^t(p)B\right) \sim p \circ \kappa + \sum_{\substack{k \geq 1 \\ 0 < |\alpha| \leq 2k}} h^k \mu_{k,\alpha}^t D_{y,\eta}^\alpha p \circ \kappa \quad \textit{near} \quad (y_0, \eta_0),$$

where the $\mu_{k,\alpha}^t$ are smooth functions of (y, η), and

$$\sigma_t\left(B\mathrm{Op}_h^t(p)A\right) \sim p \circ \kappa^{-1} + \sum_{\substack{k \geq 1 \\ 0 < |\alpha| \leq 2k}} h^k \nu_{k,\alpha}^t D_{x,\xi}^\alpha p \circ \kappa^{-1} \quad \textit{near} \quad (x_0, \xi_0),$$

where the $\nu_{k,\alpha}^t$ are smooth functions of (x, ξ).

Remark 5.5.6 The fact that two h-pseudodifferential operators P_1 and P_2 have equivalent symbols near a point (x_0, ξ_0) is sometimes expressed by saying that P_1 and P_2 are *microlocally equivalent* near (x_0, ξ_0), and denoted by

$$P_1 \sim P_2 \quad \text{microlocally near } (x_0, \xi_0).$$

Observe that this is equivalent to the fact that $(x_0, \xi_0) \notin \mathrm{FS}((P_1 - P_2)u)$ for all $u \in \mathcal{S}'(\mathbf{R}^n)$ (recall Definition 2.9.1).

In particular, the results of the previous theorem can be rewritten as

- $AB \sim I$ microlocally near (y_0, η_0);
- $BA \sim I$ microlocally near (x_0, ξ_0);
- $A\mathrm{Op}_h^t(p)B \sim \mathrm{Op}_h^t(\widetilde{p}_1)$ microlocally near (y_0, η_0) with $\widetilde{p}_1 \sim p \circ \kappa + \sum_{\substack{k \geq 1 \\ 0 < |\alpha| \leq 2k}} h^k \mu_{k,\alpha}^t D_{y,\eta}^\alpha p \circ \kappa$ in $S_{2n}(1)$;
- $B\mathrm{Op}_h^t(p)A \sim \mathrm{Op}_h^t(\widetilde{p}_2)$ microlocally near (x_0, ξ_0) with $\widetilde{p}_2 \sim p \circ \kappa^{-1} + \sum_{\substack{k \geq 1 \\ 0 < |\alpha| \leq 2k}} h^k \nu_{k,\alpha}^t D_{x,\xi}^\alpha p \circ \kappa^{-1}$ in $S_{2n}(1)$.

Of course, the assertions (ii) and (iii) of the theorem can be just seen as particular cases of (iv) (taking $p = 1$ identically), but intuitively they tell us that A and B are somehow microlocal inverses of each other.

Proof By the results of Chapter 2, it is enough to prove the result for some fixed arbitrary value of t, e.g., $t = 0$. Then setting $P = \mathrm{Op}_h^0(p)$, a direct computation gives

$$APBu(y) = \frac{1}{(2\pi h)^n} \int e^{i(y-y')\eta/h} c(y, y', \eta) u(y') dy' d\eta$$

with

$$c(y, y', \eta) = \frac{1}{(2\pi h)^{2n}} \int e^{i[y'(\eta-\eta') + \varphi(x', \eta') - \varphi(x, \eta) + (x-x')\xi]/h} a(x, y, \eta) \\ \times p(x, \xi) b(x', y', \eta') dx\, dx'\, d\xi\, d\eta'.$$

Then writing $\varphi(x, \eta) - \varphi(x', \eta') = (x - x')\Phi_1 + (\eta - \eta')\Phi_2$ with

$$\begin{aligned} \Phi_1 &= \Phi_1(x, x', \eta) = \int_0^1 \frac{\partial \varphi}{\partial x}((1-s)x + sx', \eta) ds, \\ \Phi_2 &= \Phi_2(x', \eta, \eta') = \int_0^1 \frac{\partial \varphi}{\partial \eta}(x', (1-s)\eta + s\eta') ds, \end{aligned}$$

and making the translation $\xi \mapsto \xi + \Phi_1$ in the integral, we get

$$c(y, y', \eta) = \frac{1}{(2\pi h)^{2n}} \int e^{i[(y'-\Phi_2)(\eta-\eta')+(x-x')\xi]/h} a(x, y, \eta) p(x, \xi + \Phi_1) \times b(x', y', \eta') dx\, dx'\, d\xi\, d\eta'.$$

Next we observe that the condition (5.5.6) implies that for any η close to η_0, the application $x \mapsto \frac{\partial \varphi}{\partial \eta}(x, \eta)$ is a diffeomorphism from a neighborhood of x_0 to a neighborhood of y_0 (indeed, the inverse is given by $\pi_1 \circ \kappa_\eta$, where $\kappa_\eta(y) = \kappa(y, \eta)$ and π_1 is the projection onto the first n coordinates). As a consequence, if η and η' are close enough to η_0, the application

$$\Psi_{(\eta,\eta')} \;:\; x \mapsto \Phi_2(x, \eta, \eta')$$

defines a change of coordinates near x_0, and thus if the supports of a and b are small enough around (x_0, y_0, η_0), we can write

$$c(y, y', \eta) = \frac{1}{(2\pi h)^{2n}} \int e^{i[(y'-x')(\eta-\eta')+\langle A(x,x',\eta,\eta')(x-x'),\xi\rangle]/h} \tilde{a}(x, y, \eta, \eta') \tilde{p}(x, x', \xi, \eta) \times \tilde{b}(x', y', \eta, \eta') J(x, \eta, \eta') J(x', \eta, \eta') dx\, dx'\, d\xi\, d\eta',$$

where

$$\tilde{a}(x, y, \eta, \eta') = a(\Psi^{-1}_{(\eta,\eta')}(x), y, \eta),$$
$$\tilde{b}(x', y', \eta, \eta') = b(\Psi^{-1}_{(\eta,\eta')}(x'), y', \eta'),$$

$J(x, \eta, \eta')$ is the absolute value of the determinant of the Jacobian of $\Psi^{-1}_{(\eta,\eta')}$, and

$$\begin{aligned} A(x, x', \eta, \eta') &= \int_0^1 \left[d\Psi^{-1}_{(\eta,\eta')}((1-s)x + sx') \right] ds, \\ \tilde{p}(x, x', \xi, \eta) &= p\left(\Psi^{-1}_{(\eta,\eta')}(x), \xi + \Phi_1(\Psi^{-1}_{(\eta,\eta')}(x), \Psi^{-1}_{(\eta,\eta')}(x'), \eta) \right). \end{aligned}$$

Finally, making the change of variable $\xi \mapsto {}^tA(x, x', \eta, \eta')^{-1}\xi$ we get

$$c(y, y', \eta) = \frac{1}{(2\pi h)^{2n}} \int e^{i[(y'-x')(\eta-\eta')+(x-x')\xi]/h} \tilde{a}(x, y, \eta, \eta') q(x, x', \xi, \eta, \eta') \times \tilde{b}(x', y', \eta, \eta') \tilde{J}(x, x', \eta, \eta') dx\, dx'\, d\xi\, d\eta',$$

with

$$\widetilde{J}(x,x',\eta,\eta') = J(x,\eta,\eta')J(x',\eta,\eta')|\det A(x,x',\eta,\eta')|^{-1}$$

and

$$q(x,x',\xi,\eta,\eta')$$
$$= p\left(\Psi^{-1}_{(\eta,\eta')}(x),\, {}^tA(x,x',\eta,\eta')^{-1}\xi + \Phi_1(\Psi^{-1}_{(\eta,\eta')}(x),\Psi^{-1}_{(\eta,\eta')}(x'),\eta)\right).$$

Now we are in a situation where the stationary phase theorem (theorem 2.6.1) can be applied, and it gives us that $c \in S_{3n}(1)$ and admits the asymptotic expansion

$$c(y,y',\eta) \sim \sum_{k\geq 0}\frac{h^k}{i^k k!}(D_{x'}\cdot D_{\eta'} + D_x\cdot D_\xi)^k\left[\tilde{a}(x,y,\eta,\eta')q(x,x',\xi,\eta,\eta')\right.$$
$$\left.\times\tilde{b}(x',y',\eta,\eta')\widetilde{J}(x,x',\eta,\eta')\right]\Big|_{\substack{x=x'=y'\\ \eta'=\eta\\ \xi=0}} \quad (5.5.9)$$

in $S_{3n}(1)$. In particular,

$$c(y,y',\eta) = a(\pi_1\circ\kappa_\eta(y'),y,\eta)b(\pi_1\circ\kappa_\eta(y'),y',\eta)\widetilde{J}(y',y',\eta,\eta)$$
$$\times p\left(\pi_1\circ\kappa_\eta(y'),\frac{\partial\varphi}{\partial x}(\pi_1\circ\kappa_\eta(y'),\eta)\right) + \mathcal{O}(h)$$

and thus, by definition,

$$c(y,y',\eta) = a(\pi_1\circ\kappa_\eta(y'),y,\eta)b(\pi_1\circ\kappa_\eta(y'),y',\eta)\widetilde{J}(y',y',\eta,\eta)\,p\circ\kappa(y',\eta) + \mathcal{O}(h).$$

Now by assumption, $a(\pi_1\circ\kappa_\eta(y_0),y_0,\eta_0) = a(x_0,y_0,\eta_0)\neq 0$, and since also $\widetilde{J}\neq 0$, we can find b supported near (x_0,y_0,η_0) such that

$$\tilde{a}(y',y,\eta,\eta)\tilde{b}(y',y',\eta,\eta)\widetilde{J}(y',y',\eta,\eta) = 1 \text{ near } \{y=y'=y_0\ ,\ \eta=\eta_0\}.$$

By (5.5.9), this means that for any $p\in S_{2n}(1)$, APB is an h-pseudodifferential operator that admits a symbol $\sigma_{APB}\in S_{3n}(1)$ satisfying

$$\sigma_{APB}(y,y',\eta)\sim p\circ\kappa + \sum_{\substack{k\geq 1\\ 0<|\alpha|\leq 2k}} h^k\mu_{k,\alpha}D^\alpha_{y',\eta}p\circ\kappa(y',\eta) \quad \text{near} \quad (y_0,y_0,\eta_0),$$

where the $\mu_{k,\alpha}$ are smooth functions of (y,y',η). In particular, by the results of Chapter 2 this proves the first half of the assertions (i) and (iv) of the theorem, and taking $p=1$ this also proves (ii).

Now we see in the same way that BPA is an h-pseudodifferential operator, and by an analogous construction (but this time using also the fact that for any x the application $\eta \mapsto \nabla_x\varphi(x,\eta)$ is a local diffeomorphism) we also get an operator $\widetilde{B}$ of the same type as B such that

$$\widetilde{B}PA = \mathrm{Op}_h^t(\widetilde{p}_2)$$

with

$$\widetilde{p}_2 \sim p\circ\kappa^{-1} + \sum_{\substack{k\geq 1\\ 0<|\alpha|\leq 2k}} h^k \nu_{k,\alpha}^t D_{x,\xi}^\alpha p\circ\kappa^{-1} \quad \text{near} \quad (x_0,\xi_0)$$

where the $\nu_{k,\alpha}^t$ are smooth functions of (x,ξ). In particular, $\widetilde{B}A = I$ microlocally near (x_0,ξ_0), and as a consequence one can write

$$BPA \sim (\widetilde{B}A)BPA = \widetilde{B}(AB)PA \sim \widetilde{B}PA = \mathrm{Op}_h^t(\widetilde{p}_2),$$

where the meaning of the $\sim$'s is the same as before microlocally near (x_0,ξ_0). This proves the last part of the assertion (iv) of the theorem, and thus also (iii) by taking $p = 1$. ◇

In order to give a generalization of the previous theorem to canonical transformations that do not admit a generating function, we first need the following result:

Proposition 5.5.7 *Let κ be a symplectic transformation defined near some point (y_0,η_0) of $\mathbf{R}^{2n}$ with $\eta_0\neq 0$, and let $(x_0,\xi_0)=\kappa(y_0,\eta_0)$. Then there exists a smooth change of variables $f\ :\ y\mapsto\widetilde{y}=f(y)$ defined near y_0 such that the symplectic transformation $\widetilde{\kappa}(\widetilde{y},\widetilde{\eta}) := \kappa(f^{-1}(\widetilde{y}), {}^t df(\widetilde{y})\widetilde{\eta})$ admits a generating function near $(\widetilde{y}_0,\widetilde{\eta}_0) := (f(y_0), {}^t df(y_0)^{-1}\eta_0)$.*

Proof Denote by $\mathcal{V}_0$ a small enough open neighborhood of ξ_0 in $\mathbf{R}^n$, and consider

$$\Lambda_0 = \{(x_0,\xi)\ ;\ \xi\in\mathcal{V}_0\},$$

which is a Lagrangian submanifold of $\mathbf{R}^{2n}$, and

$$\Lambda_1 = \kappa^{-1}(\Lambda_0),$$

which is Lagrangian, too, since κ^{-1} is symplectic. We first prove a lemma:

Lemma 5.5.8 *There exists a smooth change of variables* $f : y \mapsto \widetilde{y} = f(y)$ *defined near* y_0 *such that if we denote by* $\phi_f(y,\eta) = (f(y), {}^t df(y)^{-1}\eta)$ *its associated canonical transformation, then the projection*

$$\begin{aligned} \pi_2 : \phi_f(\Lambda_1) &\rightarrow \mathbf{R}^n, \\ (\widetilde{y}, \widetilde{\eta}) &\mapsto \widetilde{\eta}, \end{aligned}$$

is a local diffeomorphism near $(\widetilde{y}_0, \widetilde{\eta}_0) := \phi_f(y_0, \eta_0)$.

Proof We start by working on the tangent spaces corresponding to the point $\rho_0 := (y_0, \eta_0)$. Denoting by k the codimension of $\pi_2(T_{\rho_0}\Lambda_1)$ in $\mathbf{R}^n$, we can find an orthogonal transformation U such that

$$U(\pi_2(T_{\rho_0}\Lambda_1)) = \{y \in \mathbf{R}^n \; ; \; y' := (y_1, \dots, y_k) = 0\}.$$

But then, writing $y'' = (y_{k+1}, \dots, y_n)$ for any point $y = (y_1, \dots, y_n) \in \mathbf{R}^n$, and $\phi_U(y,\eta) = (U(y), {}^tU^{-1}(\eta)) = (U(y), U(\eta))$ the canonical transformation associated with U, we see that the projection

$$\phi_U(T_{\rho_0}\Lambda_1) \ni (y,\eta) \mapsto (y'', \eta') \in \mathbf{R}^n$$

is an isomorphism: Indeed, if $y' = 0$ and $(y'', \eta') = 0$, then since $\phi_U(T_{\rho_0}\Lambda_1)$ is Lagrangian, we must have for all $(\widetilde{y}, \widetilde{\eta}) \in \phi_U(T_{\rho_0}\Lambda_1)$,

$$0 = \sigma((y,\eta), (\widetilde{y}, \widetilde{\eta})) = \eta\widetilde{y} - \widetilde{\eta}y = \eta''\widetilde{y}'',$$

and therefore, since $\widetilde{y}'' \in \mathbf{R}^{n-k}$ is arbitrary, one necessarily gets $\eta'' = 0$ and thus $(y, \eta) = 0$. This proves that the previous projection is injective, and thus also surjective, since the two spaces have the same dimension. As a consequence, any point (y, η) of the vector space $\phi_U(T_{\rho_0}\Lambda_1)$ is completely determined by the value of (y'', η'), which means that there exist two matrices B and C such that

$$\phi_U(T_{\rho_0}\Lambda_1) = \{(y,\eta) \in \mathbf{R}^{2n} \; ; \; y' = 0 \; , \; \eta'' = By'' + C\eta'\}.$$

Since moreover $\sigma \Big|_{\phi_U(T_{\rho_0}\Lambda_1)} = 0$ and

$$\sigma((0, y''; \eta', By'' + C\eta'), (0, \widetilde{y}''; \widetilde{\eta}', B\widetilde{y}'' + C\widetilde{\eta}')) = \langle By'' + C\eta', \widetilde{y}''\rangle - \langle y'', B\widetilde{y}'' + C\widetilde{\eta}'\rangle,$$

we get necessarily that B is symmetric and $C = 0$, and thus

$$\phi_U(T_{\rho_0}\Lambda_1) = \{y' = 0 \; , \; \eta'' = By''\}. \tag{5.5.10}$$

Before proceeding, let us observe that we can also assume without loss of generality that $y_0 = 0$ and $U(\eta_0) = (0, \dots, 0, 1)$: Indeed the first condition can be obtained just by making a translation in y, while the second one is obtained by first making a rotation in y (which also induces the same rotation in η) in such a way that $U(\eta_0)$ becomes of the form $(0, \dots, 0, \alpha)$ with $\alpha \neq 0$, and then a dilation in y (which induces the inverse dilation in η) to get $U(\eta_0) = (0, \dots, 0, 1)$.

Now choose any real symmetric $(n-k) \times (n-k)$ matrix D such that $B - D$ is invertible on $\mathbf{R}^{n-k}$, and set

$$\begin{cases} \widetilde{y}_n = y_n + \dfrac{1}{2}\langle Dy'', y''\rangle, \\ \widetilde{y}_j = y_j \text{ for } j \neq n. \end{cases}$$

Then the application $y \mapsto \widetilde{y}(y) = (\widetilde{y}_1, \dots, \widetilde{y}_n)$ defines a change of variable near 0, and we set

$$L = \left\{ (y, \eta) \; ; \; \eta = \frac{\partial}{\partial y}\widetilde{y}_n(y) \right\},$$

which is Lagrangian and contains $\rho_1 := \phi_U(y_0, \eta_0)$. Moreover,

$$T_{\rho_1} L = \{\eta' = 0, \eta'' = Dy''\},$$

and thus by (5.5.10) and the assumption on D,

$$T_{\rho_1} L \cap \phi_U(T_{\rho_0}\Lambda_1) = \{0\},$$

which means that L and $\phi_U(\Lambda_1)$ are transversal at ρ_1.

Moreover, in the new symplectic coordinates $(\widetilde{y}, \widetilde{\eta}) := (\widetilde{y}(y), {}^t d\widetilde{y}(y)^{-1}\eta) = \phi_{\widetilde{y}}(y, \eta)$, one has

$$\phi_{\widetilde{y}}(L) = \left\{ (\widetilde{y}, \widetilde{\eta}); \widetilde{\eta} = {}^t d\widetilde{y}(y)^{-1} \nabla_y \widetilde{y}_n \right\},$$

that is,

$$\phi_{\widetilde{y}}(L) = \{(\widetilde{y}, \widetilde{\eta}); \widetilde{\eta} = (0, \dots, 0, 1)\},$$

since $\nabla_y \widetilde{y}_n$ is the last column of ${}^t d\widetilde{y}(y)$. In particular, setting $\widetilde{\rho}_0 = \phi_{\widetilde{y}} \circ \phi_U(\rho_0)$, we get

$$T_{\widetilde{\rho}_0} \phi_{\widetilde{y}}(L) = \{\widetilde{\eta} = 0\},$$

and thus the previous discussion shows that $\phi_{\tilde{y}} \circ \phi_U(\Lambda_1)$ is transversal to $\{\tilde{\eta} = 0\}$ at $\tilde{\rho}_0$. As a consequence, with $f := \tilde{y} \circ U$ we see that the projection

$$\begin{array}{rccc} \pi_2 : & \phi_f(\Lambda_1) & \to & \mathbf{R}^n, \\ & (\tilde{y}, \tilde{\eta}) & \mapsto & \tilde{\eta}, \end{array}$$

is a local diffeomorphism near $\tilde{\rho}_0$. ◇

Completion of the Proof of Proposition 5.5.7 We deduce from the previous lemma that for any fixed x close enough to x_0, the projection

$$\pi_2 : \phi_f \circ \kappa^{-1}(\{(x, \xi) \; ; \; \xi \in \mathcal{V}_0\}) \ni (\tilde{y}, \tilde{\eta}) \mapsto \tilde{\eta}$$

is a local diffeomorphism, and thus so is

$$\tilde{\pi} : \mathcal{G}_\kappa := \{(\tilde{y}, \tilde{\eta}, x, \xi) \; ; \; (x, \xi) = \kappa \circ \phi_f^{-1}(\tilde{y}, \tilde{\eta})\} \ni (\tilde{y}, \tilde{\eta}, x, \xi) \mapsto (x, \tilde{\eta}).$$

As a consequence, there exists a smooth function ψ defined near $(x_0, \tilde{\eta}_0)$ (where we have set $\tilde{\rho}_0 = (\tilde{y}_0, \tilde{\eta}_0)$) with values in $\mathbf{R}^{2n}$ such that locally near $(x_0, \xi_0, \tilde{y}_0, \tilde{\eta}_0)$ we have the equivalence

$$(x, \xi) = \kappa \circ \phi_f^{-1}(\tilde{y}, \tilde{\eta}) \Leftrightarrow (y, \xi) = \psi(x, \tilde{\eta}). \tag{5.5.11}$$

Finally, we observe that since $\kappa \circ \phi_f^{-1}$ is symplectic, the set $\mathcal{G}_\kappa$ is a Lagrangian submanifold of $\mathbf{R}^{4n}$ respectively to the symplectic form $d\xi \wedge dx - d\tilde{\eta} \wedge d\tilde{y} = d\xi \wedge dx + d\tilde{y} \wedge d\tilde{\eta}$. As a consequence, using Proposition 5.4.4 we see that ψ is necessarily of the form

$$\psi = \left(\frac{\partial \varphi}{\partial x}, \frac{\partial \varphi}{\partial \tilde{\eta}} \right)$$

with φ smooth near $(x_0, \tilde{\eta}_0)$. By (5.5.11), this gives exactly the result of Proposition 5.5.7. ◇

Now if we have a general symplectic transformation κ defined near some point $(x_0, \xi_0) \in \mathbf{R}^{2n}$, and if we set $(y_0, \eta_0) = \kappa^{-1}(x_0, \xi_0)$, we can first make a translation in η in order to transform η_0 into $\eta_0' \neq 0$, and then (using the previous proposition) a change of the y-coordinates that makes κ admit a generating function. Then one can apply Theorem 5.5.5 in order to transform any pseudodifferential operator. Summing up all this, we can state the following result:

Theorem 5.5.9 (Egorov 2) *Let κ be a canonical transformation defined near some point $(y_0, \eta_0) = \kappa^{-1}(x_0, \xi_0) \in \mathbf{R}^{2n}$. Then there exist $\eta_0' \in \mathbf{R}^n$, a change of variables $f : y \mapsto \widetilde{y}$ defined near y_0, and a smooth function $\varphi = \varphi(x, \widetilde{\eta})$ defined near $(x_0, \widetilde{\eta}_0) := (x_0, {}^t df(y_0)^{-1}(\eta_0 - \eta_0'))$ such that for any $a = a(x, \widetilde{y}, \widetilde{\eta}; h) \in S_{3n}(1)$ supported in a small enough neighborhood of $(x_0, \widetilde{y}_0, \widetilde{\eta}_0) := (x_0, f(y_0), \widetilde{\eta}_0)$ and elliptic at $(x_0, \widetilde{y}_0, \widetilde{\eta}_0)$, there exists $b = b(x, \widetilde{y}, \widetilde{\eta}; h) \in S_{3n}(1)$ such that, denoting by A and B the operators defined as in (5.5.7) and (5.5.8), we have*

(i) $AB \sim I$ microlocally near $(\widetilde{y}_0, \widetilde{\eta}_0)$;

(ii) $BA \sim I$ microlocally near (x_0, ξ_0);

(iii) If we set

$$U_f : \varphi \mapsto \varphi \circ f$$

and

$$\widetilde{A} = e^{iy\eta_0'/h} U_f A, \quad \widetilde{B} = B U_f^{-1} e^{-iy\eta_0'/h},$$

then for any $p \in S_{2n}(1)$ and for any $t \in [0,1]$ the operator $\widetilde{A}\mathrm{Op}_h^t(p)\widetilde{B}$ is an h-pseudodifferential operator with t-symbol satisfying

$$\sigma_t\left(\widetilde{A}\mathrm{Op}_h^t(p)\widetilde{B}\right) \sim p \circ \kappa + \sum_{\substack{k \geq 1 \\ 0 < |\alpha| \leq 2k}} h^k \mu_{k,\alpha}^t D_{y,\eta}^\alpha p \circ \kappa \quad \text{near} \quad (y_0, \eta_0),$$

where the $\mu_{k,\alpha}^t$ are smooth functions of (y, η). Similarly, the operator $\widetilde{B}\mathrm{Op}_h^t(p)\widetilde{A}$ is an h-pseudodifferential operator with t-symbol satisfying

$$\sigma_t\left(\widetilde{B}\mathrm{Op}_h^t(p)\widetilde{A}\right) \sim p \circ \kappa^{-1} + \sum_{\substack{k \geq 1 \\ 0 < |\alpha| \leq 2k}} h^k \nu_{k,\alpha}^t D_{x,\xi}^\alpha p \circ \kappa^{-1} \quad \text{near} \quad (x_0, \xi_0)$$

where the $\nu_{k,\alpha}^t$ are smooth functions of (x, ξ).

A typical example of application of this theorem concerns the result on the propagation of the frequency set for the solutions (e.g., in L^2) of a partial differential equation of the form $P(x, hD_x)u = 0$, near a point (x_0, ξ_0) for which the principal symbol $p(x, \xi)$ of $P(x, hD_x)$ is real and satisfies $\nabla p(x_0, \xi_0) \neq 0$ (see also Exercise 7 of Chapter 4). We do not want to write down here all the details about it (the interested reader may consult, e.g., [GrSj]), but let us just explain the main idea: Since $\nabla p(x_0, \xi_0) \neq 0$, one can take $p(x, \xi)$ as the first momentum coordinate of a system of symplectic coordinates near (x_0, ξ_0), which means that one can find a local symplectic transformation κ such that

$$p \circ \kappa(y, \eta) = \eta_1$$

near $(y_0, \eta_0) := \kappa^{-1}(x_0, \xi_0)$. Then one can use the previous Egorov theorem to transform P microlocally into a pseudodifferential operator $\widetilde{P}$ of the type

$$\widetilde{P} = hD_{y_1} + hR,$$

where R is of order 0. Finally, using the pseudodifferential calculus of Chapter 2, one can easily construct (see, e.g., Exercise 14 of Chapter 2) an elliptic pseudodifferential operator $C = C(y, hD_y)$ such that

$$\widetilde{P}C \sim C \cdot hD_{y_1} \text{ microlocally near } (y_0, \eta_0).$$

As a consequence, the solutions u of the beginning are transformed in this way into functions $\widetilde{u}$ that satisfy

$$(y_0, \eta_0) \notin \mathrm{FS}(\partial_{y_1}\widetilde{u}).$$

The final result is then easily deduced by writing, for any $y = (y_1, y') \in \mathbf{R} \times \mathbf{R}^{n-1}$ close to $y_0 = (y_1^0, y_0')$,

$$\widetilde{u}(y) = \widetilde{u}(y_1^0, y') + \int_{y_1^0}^{y_1} \partial_{y_1}\widetilde{u}(t, y')dt,$$

and by observing that the integral curves of H_p have been changed under κ into the lines $\{(y', \eta) = \text{constant of } \mathbf{R}^{2n-1}\}$.

5.6 Exercises and Problems

1. Prove that the canonical transformations j_B and k_C introduced in Section 3.4 admit a generating function, while ℓ_j does not. Moreover, prove that the Fourier integral operators associated with these generating functions as in (5.5.7) with symbol $a = 1$ are respectively the operators J_B and K_C introduced in Section 3.4. (Hint: The corresponding generating functions are respectively $\varphi_B(y, \eta) := \langle B^{-1}x, \eta\rangle$ and $\varphi_C(x, \eta) := \langle x, \eta + Cx/2\rangle$.)

2. **R-Lagrangian Submanifolds** - On $\mathbf{C}^{2n} = \mathbf{C}^n_z \times \mathbf{C}^n_\zeta$ one considers the symplectic 2-form $\sigma_{\mathbf{R}} := \operatorname{Re}(d\zeta \wedge dz)$, that is, for $u = (u_z, u_\zeta)$ and $v = (v_z, v_\zeta) \in \mathbf{C}^{2n}$,

$$\sigma_{\mathbf{R}}(u, v) = \operatorname{Re}(u_\zeta \cdot v_z - u_z \cdot v_\zeta).$$

Then a (C^∞-) submanifold Λ of $\mathbf{C}^{2n}$ is called **R**-Lagrangian if its dimension on **R** is $2n$ and $\sigma_{\mathbf{R}}|_\Lambda = 0$. Assuming that the projection $\Lambda \ni (z, \zeta) \mapsto z \in \mathbf{C}^n$ is a local diffeomorphism, prove that Λ is **R**-Lagrangian if and only if it is locally described by an equation of the type

$$\zeta = \frac{\partial}{\partial z}\phi_{\mathbf{R}}(z),$$

where $\phi_{\mathbf{R}}$ is a smooth *real-valued* function of $z \in \mathbf{C}^n$ and

$$\frac{\partial}{\partial z} := \frac{1}{2}\left(\frac{\partial}{\partial \operatorname{Re} z} - i\frac{\partial}{\partial \operatorname{Im} z}\right).$$

(Hint: Write $\sigma_{\mathbf{R}} = d\operatorname{Re}\zeta \wedge d\operatorname{Re} z - d\operatorname{Im}\zeta \wedge d\operatorname{Im} z$ and, working in $\mathbf{R}^{4n}$, mimic the proof of Proposition 5.4.4.)

3. **I-Lagrangian Submanifolds** - In the same spirit, a smooth submanifold Λ of $\mathbf{C}^{2n}$ is called **I**-Lagrangian if its dimension on **R** is $2n$ and $\sigma_{\mathbf{I}}|_\Lambda = 0$, where $\sigma_{\mathbf{I}} := \operatorname{Im}(d\zeta \wedge dz)$. Assuming again that the projection $\Lambda \ni (z, \zeta) \mapsto z \in \mathbf{C}^n$ is a local diffeomorphism, prove that Λ is **I**-Lagrangian if and only if it is locally described by an equation of the type

$$\zeta = i\frac{\partial}{\partial z}\phi_{\mathbf{I}}(z),$$

where $\phi_{\mathbf{I}}$ is a smooth real-valued function of z.

4. **C-Lagrangian Submanifolds** - Now, a smooth submanifold Λ of $\mathbf{C}^{2n}$ is called **C**-Lagrangian if its dimension on **R** is $2n$ and $\sigma_{\mathbf{C}}|_\Lambda = 0$, where $\sigma_{\mathbf{C}} := d\zeta \wedge dz = \sigma_{\mathbf{R}} + i\sigma_{\mathbf{I}}$. Still assuming that the projection $\Lambda \ni (z, \zeta) \mapsto z \in \mathbf{C}^n$ is a local diffeomorphism, prove that Λ is **C**-Lagrangian if and only if it is locally described by an equation of the type

$$\zeta = \frac{\partial \phi}{\partial z}(z),$$

where ϕ is a *holomorphic* function of z (in particular, Λ is a holomorphic submanifold of $\mathbf{C}^{2n}$). What is the relation between ϕ and the functions $\phi_{\mathbf{R}}$ and $\phi_{\mathbf{I}}$ found in the two previous exercises? (Hint: Observe that Λ is **C**-Lagrangian if and only if it is both **R**-Lagrangian and **I**-Lagrangian; apply the two previous exercises, and set $\phi := (\phi_{\mathbf{R}} + i\phi_{\mathbf{I}})/2$.)

(Note: The terminology used in the three previous exercises comes from [Scha].)

5. **Positive (respectively Negative) Submanifolds** - A smooth submanifold Λ of $\mathbf{C}^{2n}$ is called *positive* (respectively *negative*) at some point $\rho \in \Lambda$ if the (real-valued) quadratic form $Q : u \mapsto \frac{1}{i}\sigma_{\mathbf{C}}(u, \overline{u})$ (where $\sigma_{\mathbf{C}} = d\zeta \wedge dz$ as in Exercise 4) is positive definite (respectively negative definite) on the tangent space $T_\rho\Lambda$ of Λ at ρ.

   (i) Prove that if Λ is positive or negative at some point, then its dimension on **R** is at most $2n$.

   (ii) Similarly, one says that Λ is *nonnegative* (respectively *nonpositive*) at ρ if Q is nonnegative (respectively nonpositive) on $T_\rho\Lambda$. prove that if Λ_1 and Λ_2 are two submanifolds of $\mathbf{C}^{2n}$ such that Λ_1 is negative and Λ_2 is nonnegative, then Λ_1 and Λ_2 are transversal to each other.

   (iii) Writing $u = (u_z, u_\zeta)$, establish the formula
$$\frac{1}{i}\sigma_{\mathbf{C}}(u, \overline{u}) = 2\mathrm{Im}\,(u_\zeta \cdot \overline{u}_z).$$

   (iv) Deduce from (ii)–(iii) that if Λ is positive or negative at $\rho = (\rho_z, \rho_\zeta)$, then it is necessarily transversal to the Lagrangian planes $\{z = \rho_z\}$ and $\{\zeta = \rho_\zeta\}$ at ρ. In particular, the two projections $\Lambda \ni (z, \zeta) \mapsto z \in \mathbf{C}^n$ and $\Lambda \ni (z, \zeta) \mapsto \zeta \in \mathbf{C}^n$ are local immersions near ρ.

   (v) Assume now that Λ is **C**-Lagrangian in the sense of Exercise 4. Then prove that it is positive (respectively negative) at some point ρ if and only if it can be described near ρ by an equation of the type $\zeta = \frac{\partial \phi}{\partial z}(z)$, where ϕ is a holomorphic function of z such that the real symmetric matrix $\left(\mathrm{Im}\,\frac{\partial^2 \phi}{\partial z_j \partial z_k}\right)_{1 \leq j,k \leq n}$ is positive definite (respectively negative definite). State a similar result for nonnegative (respectively nonpositive) Lagrangian submanifolds).

6. **Complex Canonical Transformations** - Let $\mathcal{U}$ and $\mathcal{V}$ be two open subsets of $\mathbf{C}^{2n}$. Then a holomorphic diffeomorphism $\kappa : \mathcal{U} \rightarrow \mathcal{V}$ is called a *complex canonical transformation* (or equivalently a *complex symplectic transformation*) on $\mathcal{U}$ if it satisfies $\kappa^*\sigma_{\mathbf{C}} = \sigma_{\mathbf{C}}$ on $\mathcal{U}$, where $\sigma_{\mathbf{C}} = d\zeta \wedge dz$ as in Exercise 4, and in the sense that for all $\rho \in \mathcal{U}$ and $u, v \in \mathbf{C}^{2n}$ one has

$$\sigma_{\mathbf{C}}(d\kappa(\rho) \cdot u, d\kappa(\rho) \cdot v) = \sigma_{\mathbf{C}}(u, v)$$

(here $d\kappa$ stands for the holomorphic differential of κ).

(i) Prove that if κ is a complex canonical transformation on $\mathcal{U}$ and $\Lambda \subset \mathcal{U}$ is $\mathbf{C}$-Lagrangian (in the sense of Exercise 4 above), then so is $\kappa(\Lambda)$.

(ii) Prove that the following transformations are complex canonical on $\mathbf{C}^{2n}$:

(a) $(z, \zeta) \mapsto (Bz, {}^tB^{-1}\zeta)$, where B is an invertible $n \times n$ complex matrix;

(b) $(z, \zeta) \mapsto (z + C\zeta, \zeta)$, where C is a complex $n \times n$ symmetric matrix (i.e. ${}^tC = C$);

(c) $(z, \zeta) \mapsto ((-\zeta_1, z'), (z_1, \zeta'))$, where $z' := (z_2, \ldots, z_n)$ and $\zeta' := (\zeta_2, \ldots, \zeta_n)$.

(iii) Let $p = p(z, \zeta)$ be a holomorphic function on an open subset $\mathcal{U}$ of $\mathbf{C}^{2n}$. Then as in the real domain one defines the Hamilton field of p by the identity

$$\sigma_{\mathbf{C}}(u, H_p(z, \zeta)) = dp(z, \zeta) \cdot u$$

(see Section 5.3 and Exercise 4 above). One also defines the Hamilton flow $\exp tH_p(z, \zeta)$ (t real) as in Definition 4.3.2. Then prove that for any open subset $\mathcal{U}' \subset\subset \mathcal{U}$ and for any $t \in \mathbf{R}$ such that $\bigcup_{s \in [0,t]} \exp sH_p(\mathcal{U}') \subset \mathcal{U}$, the application

$$\mathcal{U}' \ni (z, \zeta) \mapsto \exp tH_p(z, \zeta)$$

is a complex canonical transformation. (Hint: Mimic, in the holomorphic context, the proof of Proposition 5.3.2.)

7. **Canonical Transformation Associated with the Bargman Transform** - We slightly extend the definition of the Bargman transform introduced in Section 3 (see (3.1.2)) by setting

$$\widetilde{T}_\mu u(z;h) = \int e^{-\mu(z-y)^2/2h} u(y)dy,$$

where $\mu > 0$ is arbitrary. Prove the existence of a (unique) linear complex canonical transformation κ_μ (in the sense of Exercise 6) such that for any linear function $L = L(x,\xi)$ and for any $u \in \mathcal{S}'(\mathbf{R}^n)$ one has

$$\widetilde{T}_\mu(L(x,hD_x)u)(z;h) = [(L \circ \kappa_\mu^{-1})(z,hD_z)]\widetilde{T}_\mu u(z;h).$$

More generally, prove that for any differential operator $P(x,hD_x)$ with polynomial coefficients one has

$$\widetilde{T}_\mu(P(x,hD_x)u)(z;h) = \widetilde{P}(z,hD_z;h)\widetilde{T}_\mu u(z;h),$$

where the principal symbol of $\widetilde{P}$ is $\widetilde{p} := p \circ \kappa_\mu^{-1}$. (Hint: One obtains $\kappa_\mu(z,\zeta) = (z - i\mu^{-1}\zeta, \zeta)$.) Here κ_μ is usually called the *canonical transformation associated with* $\widetilde{T}_\mu$.

8. **Propagation of the Positivity** - Let Λ be a $\mathbf{C}$-Lagrangian submanifold of $\mathbf{C}^{2n}$ in the sense of Exercise 4, and assume that there exists $\rho \in \Lambda \cap \mathbf{R}^{2n}$ such that Λ is positive at ρ in the sense of Exercise 5. Moreover, assume that there exists a complex canonical transformation κ (in the sense of Exercise 6) defined on a complex domain containing $\mathbf{R}^{2n}$, such that $\kappa(\mathbf{R}^{2n}) \subset \mathbf{R}^{2n}$ and $\kappa(\rho) \in \Lambda$. Then prove that Λ is positive at $\kappa(\rho)$, too. (Hint: Observe that if $u \in T_{\kappa(\rho)}\Lambda$, then $u = d\kappa(\rho)\cdot v$ with $v \in T_\rho\Lambda$ and $\overline{u} = \overline{d\kappa(\rho)} \cdot \overline{v} = d\kappa(\overline{\rho}) \cdot \overline{v} = d\kappa(\rho) \cdot \overline{v}$.)

9. **Suppression of Caustics** - Let Λ be a $\mathbf{C}$-Lagrangian submanifold of $\mathbf{C}^{2n}$ (see Exercise 4) such that $\Lambda \cap \mathbf{R}^{2n}$ is nonempty and connected, and let $p = p(x,\xi)$ be a real-valued analytic function on $\mathbf{R}^{2n}$ such that $p\,|_{\Lambda\cap\mathbf{R}^{2n}} = 0$. We assume that there exists $\rho_0 = (\rho_z^0, \rho_\zeta^0) \in \Lambda \cap \mathbf{R}^{2n}$ such that near ρ_0, Λ is given by an equation of the type $\zeta = \partial_z\varphi(z)$ with φ holomorphic and $\operatorname{Im}\operatorname{Hess}\varphi(\rho_z^0) \geq 0$. In particular, φ satisfies near ρ_z^0 the eikonal equation

$$p(z, \partial_z\varphi(z)) = 0, \tag{5.6.1}$$

and by analyticity this equation remains valid on any connected domain of $\Lambda \cap \mathbf{R}^{2n}$ where φ can be extended as an analytic function. However,

φ ceases to exist at those points of $\Lambda \cap \mathbf{R}^{2n}$ where the projection $\Pi_z \mid_\Lambda : \Lambda \ni (z,\zeta) \mapsto z$ fails to be a local diffeomorphism. Such points are called *caustic* points, and the set they form is called the *caustic set* of Λ. In particular, the solution φ of (5.6.1) cannot be extended near this set. To somehow overcome this problem, a technique consists in using the canonical transformation κ_μ, which was introduced in Exercise 7 (and in this case μ is usually taken large), in the following way:

(i) Prove that $\kappa_\mu^{-1}(\{(0,\zeta)\ ;\ \zeta \in \mathbf{C}^n\})$ is a negative submanifold of $\mathbf{C}^{2n}$ (in the sense of Exercise 5), and that Λ is nonnegative at ρ_0. (Hint: Use Exercise 5 (v).)

(ii) Differentiating the identity $p(z, \partial_z\varphi(z)) = 0$ with respect to z, prove that the vector field H_p is tangent to Λ near ρ_0, and thus everywhere on $\Lambda \cap \mathbf{R}^{2n}$, by the unique analytic continuation theorem.

(iii) For any $\rho \in \mathbf{C}^{2n}$ we denote by $T_\rho^\pm \geq 0$ the largest real numbers such that $\exp tH_p(\rho)$ is well-defined for $t \in (-T_\rho^-, T_\rho^+)$ (in particular, $T_\rho^\pm = 0$ if p is not defined at ρ). Deduce from (ii) that for any $\rho \in \Lambda$ and $t \in (-T_\rho^-, T_\rho^+)$ one has $\exp tH_p(\rho) \in \Lambda$.

(iv) We set

$$\Lambda_0 := \bigcup_{\rho\in\Lambda} \left\{\exp tH_p(\rho)\ ;\ t \in (-T_\rho^-, T_\rho^+)\right\}.$$

Prove that Λ_0 is a $\mathbf{C}$-Lagrangian submanifold of $\mathbf{C}^{2n}$ that contains $\Lambda \cap \mathbf{R}^{2n}$ and on which p vanishes identically. (Hint: Use (ii) and Exercise 6.)

(v) Deduce from (i) that Λ_0 is nonnegative at $\exp tH_p(\rho_0)$ for any $t \in (-T_{\rho_0}^-, T_{\rho_0}^+)$. (Hint: Proceed as in Exercise 8.)

(vi) Deduce from (i) and (v) that there exist a complex neighborhood Ω of the curve $\{\kappa_\mu(\exp tH_p(\rho_0))\ ;\ t \in (-T_{\rho_0}^-, T_{\rho_0}^+)\}$ and a function ϕ holomorphic on $\Pi_z(\Omega)$ such that

$$\kappa_\mu(\Lambda_0) \cap \Omega = \left\{(z,\zeta) \in \Omega\ ;\ \zeta = \frac{\partial\phi}{\partial z}(z)\right\}.$$

(vii) Give a conclusion concerning the "modified" eikonal equation

$$p(z + i\mu^{-1}\partial_z\phi(z), \partial_z\phi(z)) = 0$$

near the curve $\{\kappa_\mu(\exp tH_p(\rho_0))\ ;\ t \in (-T_{\rho_0}^-, T_{\rho_0}^+)\}$.

(viii) **Application to the Schrödinger Operator** - Consider the case $p(x,\xi) = \xi^2 + V(x) - E$ (which corresponds to the study of the Schrödinger operator near the energy level E). Assume that V is analytic and uniformly bounded from below on $\mathbf{R}^n$, and suppose there exists φ real-valued and analytic near some $x_0 \in \mathbf{R}^n$ such that $|\nabla\varphi(x)|^2 = E - V(x)$ near x_0. Denote by $(x(t), \xi(t)) = \exp tH_p(x_0, \nabla\varphi(x_0))$ ($t \in \mathbf{R}$) the classical trajectory of energy E starting at $(x_0, \nabla\varphi(x_0))$.

(viii-a) Deduce from (vi) that for any $\varepsilon > 0$ there exists $\phi_\varepsilon = \phi_\varepsilon(z)$ holomorphic near $\{x(t) - i\varepsilon\xi(t)\ ;\ t \in \mathbf{R}\}$ such that $\nabla\phi_\varepsilon(x(t) - i\varepsilon\xi(t))$ is real for all t and

$$|\nabla\phi_\varepsilon(x - i\xi)|^2 + V(x - i\varepsilon\xi + i\varepsilon\nabla\phi_\varepsilon(x - i\varepsilon\xi)) = E$$

in a neighborhood of $\{(x(t), \xi(t))\ ;\ t \in \mathbf{R}\}$.

(viii-b) In the case where the potential V is polynomial, prove the existence of an elliptic symbol $a_\varepsilon(z,h) \sim \sum_{j\geq 0} h^j a_{\varepsilon,j}(z)$ holomorphic with respect to z such that

$$\left(-h^2\Delta_z + V(z + i\varepsilon hD_z) - E\right)\left(a_\varepsilon e^{i\phi_\varepsilon/h})\right) \sim 0$$

for z in a complex neighborhood of $\{x(t) - i\varepsilon\xi(t)\ ;\ t \in \mathbf{R}\}$. (Hint: Proceed in a similar way as in Exercise 7 of Chapter 2.) Note: Because of its links with the Bargman transform (see Exercise 7), such a solution $a_\varepsilon e^{i\phi_\varepsilon/h}$ is called a *microlocal WKB solution* (indeed, this method would also give an approximate solution v of $\widetilde{P}v = 0$, where $\widetilde{P}$ is deduced from $P = -h^2\Delta + V - E$ as in Exercise 7).

Chapter 6

Appendix: List of Formulae

Here we present a list of useful formulae encountered in the main text, to which we have added formulae coming from the related theory of microlocal analysis (not necessarily semiclassical). Since this is merely a list, we use either the notation of the main text, or anyway intuitive enough notation in order to avoid making the text heavier by introducing it each time. In any case, references to the main text are given, where the reader will get both the specification of the notation and a proof of the formula. The formulae for nonsemiclassical calculus are obtained by formally replacing h by 1 in the semiclassical formulae. The meaning of the asymptotics depends on the calculus one considers (e.g., modulo $\mathcal{O}(\langle\xi\rangle^{-\infty})$ as $|\xi| \to \infty$ for symbols decaying more and more as they are more and more differentiated with respect to ξ).

6.1 Change of Quantization

6.1.1 Semiclassical Calculus

$\mathrm{Op}_h^t(b_t) = \mathrm{Op}_h^{t'}(b_{t'})$ with (see (2.7.6))

$$b_{t'}(x,\xi) = e^{ih(t-t')D_xD_\xi} b_t(x,\xi) \sim \sum_{\alpha\in\mathbf{N}^n} \frac{(t-t')^{|\alpha|}h^{|\alpha|}}{i^{|\alpha|}\alpha!} \partial_x^\alpha \partial_\xi^\alpha b_t(x,\xi).$$

6.1.2 Nonsemiclassical Calculus

$\mathrm{Op}_1^t(b_t) = \mathrm{Op}_1^{t'}(b_{t'})$ with

$$b_{t'}(x,\xi) = e^{i(t-t')D_xD_\xi} b_t(x,\xi) \sim \sum_{\alpha\in\mathbf{N}^n} \frac{(t-t')^{|\alpha|}}{i^{|\alpha|}\alpha!} \partial_x^\alpha \partial_\xi^\alpha b_t(x,\xi).$$

6.2 Composition Formulae

6.2.1 Semiclassical Calculus

General t-Quantization (see Theorem 2.7.4):

$$\begin{aligned} a \#^t b\,(x,\xi;h) &= e^{ih[D_\eta D_v - D_u D_\xi]}\Big[a((1-t)x+tu,\eta)b(tx+(1-t)v,\xi)\Big]_{\substack{u=v=x\\ \eta=\xi}} \\ &\sim \textstyle\sum_{k\geq 0} \frac{h^k}{i^k k!}(\partial_\eta\partial_v - \partial_\xi\partial_u)^k \Big[a((1-t)x+tu,\eta)b(tx+(1-t)v,\xi)\Big]_{\substack{u=v=x\\ \eta=\xi}}. \end{aligned}$$

Standard Quantization (see (2.7.11)):

$$a\#b\,(x,\xi;h) = e^{ihD_\eta D_y} a(x,\eta)b(y,\xi)\Big|_{\substack{y=x\\ \eta=\xi}} \sim \sum_\alpha \frac{h^{|\alpha|}}{i^{|\alpha|}\alpha!} \partial_\xi^\alpha a(x,\xi)\,\partial_x^\alpha b(x,\xi).$$

Weyl-Quantization (see (2.7.12)):

$$\begin{aligned} a\#^W b\,(x,\xi;h) &= e^{ih[D_\eta D_x - D_y D_\xi]} a(y,\eta)b(x,\xi)\Big|_{\substack{y=x\\ \eta=\xi}} \\ &\sim \sum_{\alpha,\beta} \frac{h^{|\alpha+\beta|}(-1)^{|\alpha|}}{(2i)^{|\alpha+\beta|}\alpha!\beta!} (\partial_x^\alpha\partial_\xi^\beta a(x,\xi))(\partial_\xi^\alpha\partial_x^\beta b(x,\xi)) \\ &= a(x,\xi)b(x,\xi) + \frac{h}{2i}\{a,b\}(x,\xi) + \mathcal{O}(h^2). \end{aligned}$$

6.2.2 Nonsemiclassical Calculus

General t-Quantization:

$$\begin{aligned} a \#^t b\,(x,\xi) &= e^{i[D_\eta D_v - D_u D_\xi]}\Big[a((1-t)x+tu,\eta)b(tx+(1-t)v,\xi)\Big]_{\substack{u=v=x\\ \eta=\xi}} \\ &\sim \sum_{k\geq 0} \frac{1}{i^k k!}(\partial_\eta\partial_v - \partial_\xi\partial_u)^k \Big[a((1-t)x+tu,\eta)b(tx+(1-t)v,\xi)\Big]_{\substack{u=v=x\\ \eta=\xi}}. \end{aligned}$$

Standard Quantization:

$$a\#b\,(x,\xi) = e^{iD_\eta D_y} a(x,\eta)b(y,\xi)\Big|_{\substack{y=x\\ \eta=\xi}} \sim \sum_\alpha \frac{1}{i^{|\alpha|}\alpha!}\partial_\xi^\alpha a(x,\xi)\,\partial_x^\alpha b(x,\xi).$$

Weyl-Quantization:

$$\begin{aligned} a\#^W b\,(x,\xi) &= e^{i[D_\eta D_x - D_y D_\xi]} a(y,\eta)b(x,\xi)\Big|_{\substack{y=x\\ \eta=\xi}} \\ &\sim \sum_{\alpha,\beta} \frac{(-1)^{|\alpha|}}{(2i)^{|\alpha+\beta|}\alpha!\beta!}(\partial_x^\alpha\partial_\xi^\beta a(x,\xi))(\partial_\xi^\alpha\partial_x^\beta b(x,\xi)) \\ &= a(x,\xi)b(x,\xi) + \frac{1}{2i}\{a,b\}(x,\xi) + \text{lower} - \text{order terms.} \end{aligned}$$

6.3 Symbol of the Adjoint and Transpose Operator

6.3.1 Semiclassical Calculus

For any $t \in [0,1]$, the adjoint A^* of the semiclassical pseudodifferential operator A satisfies (see Remark 2.5.7)

$$\sigma_{1-t}(A^*) = \overline{\sigma_t(A)}.$$

In particular, using (2.7.6), the t-symbol A^* is related to that of A by

$$\sigma_t(A^*)(x,\xi) = e^{ih(1-2t)D_x D_\xi}\overline{\sigma_t(A)(x,\xi)} \sim \sum_\alpha \frac{(1-2t)^{|\alpha|}h^{|\alpha|}}{i^{|\alpha|}\alpha!}\partial_x^\alpha\partial_\xi^\alpha\overline{\sigma_t(A)(x,\xi)}.$$

Similarly, the transpose $\tilde{A}$ of A (defined by $\left\langle \tilde{A}u, \overline{v}\right\rangle_{L^2} = \langle Av, \overline{u}\rangle_{L^2}$) satisfies

$$\sigma_{1-t}(\tilde{A})(x,\xi) = \sigma_t(A)(x,-\xi)$$

and thus

$$\begin{aligned} \sigma_t(\tilde{A})(x,\xi) &= e^{ih(1-2t)D_x D_\xi}\left[\sigma_t(A)(x,-\xi)\right] \\ &\sim \sum_\alpha \frac{(2t-1)^{|\alpha|}h^{|\alpha|}}{i^{|\alpha|}\alpha!}\left[\partial_x^\alpha\partial_\xi^\alpha\sigma_t(A)\right](x,-\xi). \end{aligned}$$

6.3.2 Nonsemiclassical Calculus

For the adjoint A^* of A:

$$\sigma_t(A^*)(x,\xi) = e^{i(1-2t)D_xD_\xi}\overline{\sigma_t(A)(x,\xi)} \sim \sum_\alpha \frac{(1-2t)^{|\alpha|}}{i^{|\alpha|}\alpha!}\partial_x^\alpha\partial_\xi^\alpha\overline{\sigma_t(A)(x,\xi)}.$$

For the transpose $\widetilde{A}$ of A:

$$\begin{aligned}\sigma_t(\widetilde{A})(x,\xi) &= e^{i(1-2t)D_xD_\xi}\left[\sigma_t(A)(x,-\xi)\right] \\ &\sim \sum_\alpha \frac{(2t-1)^{|\alpha|}}{i^{|\alpha|}\alpha!}\left[\partial_x^\alpha\partial_\xi^\alpha\sigma_t(A)\right](x,-\xi).\end{aligned}$$

6.4 Symbol of a Commutator

For any t-quantization in the semiclassical calculus (see Remark 2.7.6)

$$\sigma_t([A,B]) = \frac{h}{i}\{a,b\} + \mathcal{O}(h^2).$$

(In the nonsemiclassical case, $\sigma_t([A,B]) = \frac{1}{i}\{a,b\}$ + lower-order terms.)

6.5 Functional Calculus

For $f \in C_0^\infty(\mathbf{R})$ and A self-adjoint one has the Helffer–Sjöstrand formula (see (2.10.2))

$$f(A) = -\frac{1}{\pi}\int_{\mathbf{R}^2}\overline{\partial}\mathcal{A}_f(\mu,\nu)(\mu+i\nu-A)^{-1}d\mu d\nu$$

where $\mathcal{A}_f$ is any almost-analytic extension of f (in particular, $\mathcal{A}_f(\mu,0) = f(\mu)$ and $\overline{\partial}\mathcal{A}_f(\mu,\nu) := \frac{1}{2}(\partial_\mu + i\partial_\nu)\mathcal{A}_f(\mu,\nu) = \mathcal{O}(|\nu|^\infty)$ as $\nu \to 0$).

6.6 Fourier–Bros–Iagolnitzer Transform

Let A be an $n \times n$ real symmetric positive definite matrix. The FBI transform T_A associated with A is defined on $\mathcal{S}'(\mathbf{R}^n)$ by (see (3.1.1) and (3.4.1))

$$T_Au(x,\xi;h) = 2^{-\frac{n}{2}}(\pi h)^{-\frac{3n}{4}}(\det A)^{\frac{1}{4}}\int e^{i(x-y)\xi/h-\langle A(x-y),x-y\rangle/2h}u(y)dy,$$

and it is an isometry from $L^2(\mathbf{R}^n)$ to $L^2(\mathbf{R}^{2n})$.

6.7 Microlocal Exponential Estimates

For any symbol $q \in S_{4n}(1)$ and any real-valued $\psi \in C^\infty(\mathbf{R}^{2n})$, bounded together with all its derivatives, one has the a priori estimate (see Theorem 3.5.1 and Exercise 9 of Chapter 3)

$$\left\langle q(x,\xi,hD_x,hD_\xi)e^{\psi/h}T_Au, e^{\psi/h}T_Av\right\rangle_{L^2}$$

$$= \left\langle (q(x,\xi,\xi - A\partial_\xi\psi, A^{-1}\partial_x\psi)e^{\psi/h}Tu, e^{\psi/h}Tv\right\rangle_{L^2} + \mathcal{O}(h)\|e^{\psi/h}Tu\|\cdot\|e^{\psi/h}Tv\|.$$

In particular, if $p \in S_{2n}(1)$ admits a convenient holomorphic extension, one gets (see Corollary 3.5.3 and Exercise 9 of Chapter 3)

$$\left\langle e^{\psi/h}T_Ap(x,hD_x)u, e^{\psi/h}T_Av\right\rangle_{L^2}$$

$$= \left\langle (p_\psi(x,\xi)e^{\psi/h}Tu, e^{\psi/h}Tv\right\rangle_{L^2} + \mathcal{O}(h)\|e^{\psi/h}Tu\|\cdot\|e^{\psi/h}Tv\|$$

with

$$p_\psi(x,\xi) = p(x - A^{-1}\partial_x\psi - i\partial_\xi\psi, \xi - A\partial_\xi\psi + i\partial_x\psi).$$

6.8 Agmon Estimates

For any $h > 0$, $V \in L^\infty(\mathbf{R}^n)$ real-valued, $E \in \mathbf{R}$, $u \in H^1(\mathbf{R}^n)$, and φ real-valued and Lipschitz on $\mathbf{R}^n$, one has (see Exercise 8 of Chapter 3)

$$\mathrm{Re}\left\langle e^{\varphi/h}(-h^2\Delta + V - E)u, e^{\varphi/h}u\right\rangle$$

$$= \|h\nabla(e^{\varphi/h}u)\|^2 + \left\langle (V(x) - E - |\nabla\varphi(x)|^2)e^{\varphi/h}u, e^{\varphi/h}u\right\rangle.$$

Bibliography

[Ag] Agmon, S., *Lectures on Exponential Decay of Solutions of Second Order Elliptic Equations, Princeton University Press*, Princeton (1982).

[AlGe] Alinhac, S., Gérard, P., *Opérateurs Pseudo-différentiels et Théorème de Nash–Moser*, Interéditions, Savoirs Actuels (Paris) (1991).

[Be1] Beals, R., A General Calculus of Pseudodifferential Operators, *Duke Math. J.* **42**, 1–42 (1975).

[Be2] Beals, R., Characterization of Pseudodifferential Operators and Applications, *Duke Math. J.* **44**, 45–57 (1977).

[Bi] Bismut, J.-M., The Witten complex and the degenerate Morse inequalities, *J. Differ. Geom.* **23**, 207–240 (1986).

[Bo] Bony, J.-M., Sur l'inégalité de Fefferman–Phong, *Séminaire EDP de l'Ecole Polytechnique*, Palaiseau (France), exp. n.III, 1998–99.

[BoHeJo] Born, M., Heisenberg, W., Jordan, P., *Zur Begruendung der Matrizenmechanik, Dokumente der Naturwissenschaft, Abt. Physik*, Bd. 2. Stuttgart: Ernst Battenberg Verlag, 135 S. (1962).

[BrCoDu] Briet, P., Combes, J.-M., Duclos, P., On the Location of Resonances for Schrödinger Operators in the Semiclassical Limit II, *Comm. Math. Phys.* **12** (2), 201–222 (1987).

[CaVa1] Calderón, A.P., Vaillancourt, R., On the Boundedness of Pseudodifferential Operators, *J. Math. Soc. Japan* **23**, 374–378 (1972).

[CaVa2] Calderón, A.P., Vaillancourt, R., A class of Bounded Pseudodifferential Operators, *Proc. Nat. Acad. Sci. USA* **69**, 1185–1187 (1972).

[ChPi] Chazarain, J., Piriou, A., *Introduction à la théorie des équations aux dérivées partielles linéaires*, Gauthier Villars, Paris (1981).

[CdV1] Colin de Verdière, Y., *Spectres de graphes*, Cours Spécialisés n.4, Paris: Société Mathématique de France (1998).

[CdV2] Colin de Verdière, Y., Une introduction a la mecanique semi-classique, *Enseign. Math.*, II. Ser. 44, No.1-2, 23–51 (1998).

[CoRo] Combescure, M., Robert, D., Semiclassical Spreading of Quantum Wave Packets and Applications near Unstable Fixed Points of the Classical Flow, *Asymptotic Analysis* **14**, 377–404 (1997).

[CFKS] Cycon, H.L., Froese, R.G., Kirsch, W., Simon, B., *Schrödinger Operators, with Applications to Quantum Mechanics and Global Geometry*, Texts and Monographs in Physics, Springer Verlag (1987).

[Da] Davies, E.B., The Functional Calculus,*J. London Math. Soc.* **52**, 166–176 (1995).

[DeB] De Broglie, L., Ondes et quanta, *Comptes Rendus Hebd. Séances Acad. Sci. Paris*, t.177, 507 (1923) [see also: Ann. d. Phys., série X, p. 2, (1925)].

[Del] Delort, J.M., *F.B.I. Transformation, Second Microlocalization and Semilinear Caustics*, Springer LNM 1522 (1992).

[DiSj] Dimassi, M., Sjöstrand, J., *Spectral Asymptotics in the Semiclassical Limit*, London Math. Soc. Lecture Notes Series 268, Cambridge University Press (1999).

[Du1] Duistermaat, J.J., *Fourier integral operators*, Lecture Notes, Courant Institute of Math. Sci., New York (1973).

[Du2] Duistermaat, J.J., Oscillatory integrals, Lagrange immersions and unfoldings of singularities. (English) *Commun. Pure Appl. Math.* **27**, 207–281 (1974).

[DuHo] Duistermaat, J.J., Hörmander, L., Fourier integral operators II,*Acta Math.* **128**, 183–269 (1972).

[Dy] Dyn'kin, E.M., An Operator Calculus Based upon the Cauchy–Green Formula, *J. Soviet Math.*, **4** (4), 329–334 (1975).

[Eg] Egorov, Yu.V., On Canonical Transformations of Pseudo-differential Operators, *Uspekhi Mat. Nauk* **25**, 235–236 (1969).

[En] Enss, V., *Introduction to Asymptotic Observables for Multi-Particle Quantum Scattering*, Lecture Notes in Math. vol. 1218, 61–92, Springer (1983).

[FeMa] Fedoriuk, M.V., Maslov, V.P., *Semiclassical Approximation in Quantum Mechanics*, Math. Phys. App. Math., Reidel, Dordrecht (1981).

[FePh1] Fefferman, C., Phong, D.H., On Positivity of Pseudodifferential Operators, *Proc. Nat. Sci. USA* **75**, 4673–4674 (1978).

[FePh2] Fefferman, C., Phong, D.H., The Uncertainty Principle and Sharp Gårding Inequalities, *Comm. Pures and Appl. Math.*, Vol. 34, 285–331 (1981).

[Fo] Folland, G.B., *Harmonic Analysis in Phase Space*, Princeton University Press (1989).

[Ga] Gårding, L., Dirichlet's Problem for Linear Elliptic Partial Differential Equations, *Math. Scand.* **1**, 55–72 (1953).

[GeP] Gérard, P., Mesures semi-classiques et ondes de Bloch, *Seminaire EDP de l'Ecole Polytechnique*, Palaiseau (France), Exposé n. XVI, 1990–91.

[GMS] Gérard, C., Martinez, A., Sjöstrand, S., A Mathematical Approach to the Effective Hamiltonian in Perturbed Periodic Problems, *Commun. Math. Phys.* **142**, n2 (1991).

[GrSj] Grigis, A., Sjöstrand, J., *Microlocal Analysis for Differential Operators, An Introduction*, London Math. Soc. Lecture Notes Series 196, Cambridge University Press (1994).

[GuSt] Guillemin, V., Sternberg, S., *Geometric Asymptotics*, Amer. Math. Soc. Survey 14 (1977).

[GRT] Guillot, J.C., Ralston, J., Trubowitz, E., Semi-Classical Asymptotics in Solid State Physics, *Commun. Math. Phys.* **116**, 401–415 (1988).

[Ha] Hanges, N., Parametrices and Propagation of Singularities for Operators with Non-Involutive Characteristics, *Indiana Univ. Math. J.* **28**, 87–97 (1979).

[He1] Helffer, B.,*Semiclassical analysis for the Schrödinger operator and applications*, Lecture Notes in Math. 1336, Springer Verlag (1988).

[He2] Helffer, B.,*Opérateurs Globalement Elliptiques*, Astérisque 112 (1984).

[HeRo] Helffer, B., Robert, D., Calcul fonctionnel par la transformation de Mellin et opérateurs admissibles, *J. Funct. Anal.* **53**, n.3, 246–268 (1983).

[HeSj1] Helffer, B., Sjöstrand, J., Multiple Wells in the Semiclassical Limit I, *Comm. Part. Diff. Eq.*, vol. 9, (4), 337–408 (1984).

[HeSj2] Helffer, B., Sjöstrand, J., *Résonances en limite semi-classique*, Mémoire Bull. Soc. Math. France, n. 24/25, tome 114 (1986).

[HeSj3] Helffer, B., Sjöstrand, J., Effet tunnel pour l'équation de Schrödinger avec champ magnétique, *Ann. Sc. Norm. Sup. di Pisa*, Ser. IV, **14**(4), 625–657 (1987).

[HeSj4] Helffer, B., Sjöstrand, J., *Semiclassical Analysis for Harper Equation III*, Mémoire Bull. Soc. Math. France, n.39, t.117, fasc.4 (1989).

[HeSj5] Helffer, B., Sjöstrand, J., Equation de Schrödinger avec champ magnétique et équation de Harper, *Springer L.N. Physics*, n. 345, 118–197 (1989).

[HeSj6] Helffer, B., Sjoestrand, J., Puits multiples en mecanique semi-classique - IV: Etude du complexe de Witten, *Comm. Part. Diff. Eq.* **10**, 245–340 (1985).

[HiSi] Hislop, P., Sigal, I.M., *Introduction to Spectral Theory*, Appl. Math. Sciences 113, Springer Verlag (1996).

[Ho1] Hörmander, L., *An Introduction to Complex Analysis in Several Variables.* North Holland–Mathematical library.

[Ho2] Hörmander, L., *The Analysis of Linear Partial Differential Operators*, Vols. I to IV. Springer Verlag (1985).

[Iv] Ivrii, V., *Microlocal Analysis and Precise Spectral Asymptotics*, Springer Verlag (1998).

[JeNa] Jensen, A., Nakamura, S., Mapping Properties of Functions of Schrödinger Operators between L^p-Spaces and Besov Spaces, *Adv. Stud. Pure Math.* **23**, 187–209 (1994).

[Ju] Jung, K., *Adiabatik und Semiklassik bei Regularität vom Gevrey-Typ*, PhD Thesis, T.U Berlin (1997).

[KK] Kawai, T., Kashiwara, Microhyperbolic Pseudodifferential Operators I, *J. Math. Soc. Japan* **27**, 359–404 (1975).

[KMSW] Klein, M., Martinez, A., Seiler, R., Wang, X., On the Born–Oppenheimer Expansion for Polyatomic Molecules, *Comm. Math. Phys.* **143**, 607–639 (1992).

[KoNi] Kohn, J.J., Nirenberg, L., An Algebra of Pseudo-differential Operators, *Comm. Pure Appl. Math.* **18**, 269–305 (1965).

[LaLi] Landau, L.D., Lifshitz, E.M., *Quantum Mechanics: Non-Relativistic Theory*, Pergamon Press, London (1958).

[Mar1] Martinez, A., Estimations sur l'effet tunnel microlocal, *Séminaire E.D.P. de l'Ecole Polytechnique* (1991–92).

[Mar2] Martinez, A., Precise Exponential Estimates in Adiabatic Theory, *J. Math. Phys.* **35** (8), 3889–3915 (1994).

[Mar3] Martinez, A., Estimates on Complex Interactions in Phase Space, *Math. Nachr.* **167**, 203–254 (1994).

[Mar4] Martinez, A., Microlocal exponential estimates and applications to tunneling, in *Microlocal Analysis and Spectral Theory*, NATO ASI Series C, Vol.490, (L. Rodino, ed.), Kluwer Acad. Publ., 349–376 (1997).

[MaSo] Martinez, A., Sordoni, V., Microlocal WKB expansions, *J. Funct. Analysis* **168**, 380–402 (1999).

[Mas] Maslov, V.P., *Théorie des Perturbations et Méthodes Asymptotiques*, Dunod, Paris (1972).

[MeSj] Melin, A., Sjöstrand, J., Fourier integral operators with Complex Valued Phase Functions, *Springer Lecture Notes in Math.*, **459**, 120–223 (1974).

[Mes] Messiah, A., *Quantum Mechanics*, North Holland (1970).

[Na1] Nakamura, S., On an Example of Phase-Space Tunneling, *Ann. Inst. H. Poincaré*, vol. 63, n. 2 (1995).

[Na2] Nakamura, S., On Martinez' Method of Phase Space Tunneling, *Rev. Math. Phys.*, vol. 7, n. 3, 431–441 (1995).

[PaUr] Paul, T., Uribe, A., The Semi-classical Trace Formula and Propagation of Wave Packets, *J. Funct. Analysis* **132**, 192–249 (1995).

[ReSi] Reed, M., B. Simon, B., Methods of Modern Mathematical Physics, Vols. I to IV, Academic Press (1972).

[Ro] Robert, D., *Autour de l'approximation semi-classique*, Birkhäuser (1987).

[SKK] Sato. M., Kawai, T., Kashiwara, M., Hyperfunctions and Pseudodifferential Equations, *Springer Lecture Notes in Math.* **287**, 265–529 (1973).

[Scha] Schapira, P., Conditions de positivité dans une variété symplectique complexe. Application à l'étude des microfonctions, *Ann. Sci. Ec. Norm. Sup.* 4ème série **14**, 121–139 (1981).

[Schr1] Schrödinger, E., Quantisierung als Eigenwertproblem, *Annalen der Physik* **79** (1926).

[Schr2] Schrödinger, E., *Collected Papers on Wave Mechanics*, London and Glasgow, Blackie and Son Limited (1928).

[Schw1] Schwartz, L., *Théorie des distributions*, Hermann, Paris (1966).

[Schw2] Schwartz, L., Théorie des distributions à valeurs vectorielles, *Ann. Inst. Fourier (Grenoble)* **7**, 1–141 (1957).

[Sh] Shubin, M.A., *Pseudodifferential Operators and Spectral Theory*, Springer series in Soviet Math., Springer (1987).

[SiSo] Sigal, I.M., Soffer, A., The N-particle Scattering Problem: Asymptotic Completeness for Short Range Systems, *Ann. Math.* **126**, 35–108 (1987).

[Si] Simon, B., Semiclassical Limit of Low Lying eigenvalues I - Non Degenerate minima, *Ann. Inst. H. Poincaré*, vol. 38, n. 3, 295–307 (1983).

[Sj1] Sjöstrand, J., *Singularités analytiques microlocales*, Astérisque 95 (1982).

[Sj2] Sjöstrand, J., Function Spaces Associated to Global I-Lagrangian Manifolds, Morimoto, M. (ed.) et al., *Structure of solutions of differential equations*, Proceedings of the Taniguchi symposium, Katata, Japan, June 26–30, 1995 and the RIMS symposium, Kyoto, Japan, July 3–7, 1995. Singapore: World Scientific. 369–423 (1996).

[St] Sternberg, S., *Lectures on Differential Geometry*, Prentice Hall (1965).

[Tat] Tataru, D., On the Fefferman–Phong Inequality and Related Problems, Preprint Northwestern University (2000).

[Tay] Taylor, M., *Pseudodifferential Operators and Spectral Theory*, Princeton Univ. Press, Princeton (1981).

[Tr] Treves, F., *Introduction to Pseudodifferential and Fourier integral operators*, Vols. 1 and 2, Plenum Press, New York (1980).

[VdW] Van der Waerden, B.L., From Matrix Mechanics and Wave Mechanics to Unified Quantum Mechanics, *Notices Am. Math. Soc.* **44**, n. 3, 323–329 (1997).

[Vo] Voros, A., The Return of the Quartic Oscillator - The complex WKB-Method, *Ann. Inst. H. Poincaré* **29**, 211–338 (1983).

[Wi] Witten, E., Supersymmetry and Morse theory, *J. Differ. Geom.* **17**, 661–692 (1982).

Index

List of Notation

Universitext (continued)

Jennings: Modern Geometry with Applications
Jones/Morris/Pearson: Abstract Algebra and Famous Impossibilities
Kac/Cheung: Quantum Calculus
Kannan/Krueger: Advanced Analysis
Kelly/Matthews: The Non-Euclidean Hyperbolic Plane
Kostrikin: Introduction to Algebra
Luecking/Rubel: Complex Analysis: A Functional Analysis Approach
MacLane/Moerdijk: Sheaves in Geometry and Logic
Marcus: Number Fields
Martinez: An Introduction to Semiclassical and Microlocal Analysis
Matsuki: Introduction to the Mori Program
McCarthy: Introduction to Arithmetical Functions
Meyer: Essential Mathematics for Applied Fields
Mines/Richman/Ruitenburg: A Course in Constructive Algebra
Moise: Introductory Problems Course in Analysis and Topology
Morris: Introduction to Game Theory
Poizat: A Course In Model Theory: An Introduction to Contemporary Mathematical Logic
Polster: A Geometrical Picture Book
Porter/Woods: Extensions and Absolutes of Hausdorff Spaces
Radjavi/Rosenthal: Simultaneous Triangularization
Ramsay/Richtmyer: Introduction to Hyperbolic Geometry
Reisel: Elementary Theory of Metric Spaces
Ribenboim: Classical Theory of Algebraic Numbers
Rickart: Natural Function Algebras
Rotman: Galois Theory
Rubel/Colliander: Entire and Meromorphic Functions
Sagan: Space-Filling Curves
Samelson: Notes on Lie Algebras
Schiff: Normal Families
Shapiro: Composition Operators and Classical Function Theory
Simonnet: Measures and Probability
Smith: Power Series From a Computational Point of View
Smith/Kahanpää/Kekäläinen/Traves: An Invitation to Algebraic Geometry
Smorynski: Self-Reference and Modal Logic
Stillwell: Geometry of Surfaces
Stroock: An Introduction to the Theory of Large Deviations
Sunder: An Invitation to von Neumann Algebras
Tondeur: Foliations on Riemannian Manifolds
Toth: Finite Möbius Groups, Minimal Immersions of Spheres, and Moduli
Wong: Weyl Transforms
Zhang: Matrix Theory: Basic Results and Techniques
Zong: Sphere Packings
Zong: Strange Phenomena in Convex and Discrete Geometry

GPSR Compliance
The European Union's (EU) General Product Safety Regulation (GPSR) is a set of rules that requires consumer products to be safe and our obligations to ensure this.

If you have any concerns about our products, you can contact us on

ProductSafety@springernature.com

In case Publisher is established outside the EU, the EU authorized representative is:

Springer Nature Customer Service Center GmbH
Europaplatz 3
69115 Heidelberg, Germany

www.ingramcontent.com/pod-product-compliance
Ingram Content Group UK Ltd.
Pitfield, Milton Keynes, MK11 3LW, UK
UKHW020123200726
13856UKWH00002B/694

* 9 7 8 1 4 7 5 7 4 4 9 6 5 *